SpringerBriefs in Applied Sciences and Technology

SpringerBriefs present concise summaries of cutting-edge research and practical applications across a wide spectrum of fields. Featuring compact volumes of 50 to 125 pages, the series covers a range of content from professional to academic.

Typical publications can be:

- A timely report of state-of-the art methods
- An introduction to or a manual for the application of mathematical or computer techniques
- A bridge between new research results, as published in journal articles
- A snapshot of a hot or emerging topic
- An in-depth case study
- A presentation of core concepts that students must understand in order to make independent contributions

SpringerBriefs are characterized by fast, global electronic dissemination, standard publishing contracts, standardized manuscript preparation and formatting guidelines, and expedited production schedules.

On the one hand, **SpringerBriefs in Applied Sciences and Technology** are devoted to the publication of fundamentals and applications within the different classical engineering disciplines as well as in interdisciplinary fields that recently emerged between these areas. On the other hand, as the boundary separating fundamental research and applied technology is more and more dissolving, this series is particularly open to trans-disciplinary topics between fundamental science and engineering.

Indexed by EI-Compendex, SCOPUS and Springerlink.

Aleksandr I. Potekaev · Victor E. Gromov ·
Yurii F. Ivanov · Sergey A. Nevskii ·
Sergey V. Konovalov

High-Speed Steel After Tempering and Electron Beam Processing

Structure and Properties

Springer

Aleksandr I. Potekaev
Laboratory of Advanced Technologies
Faculty of Physics, Tomsk Scientific Center SB RAS
National Research Tomsk State University
Tomsk, Russia

Yurii F. Ivanov
Plasma Emission Electronics Laboratory
Institute of High Current Electronics SB RAS
Tomsk, Russia

Sergey V. Konovalov
Federal State Budgetary Educational Institution of Higher Education "Siberian State Industrial University"
Novokuznetsk, Russia

Victor E. Gromov
Federal State Budgetary Educational Institution of Higher Education "Siberian State Industrial University"
Novokuznetsk, Russia

Sergey A. Nevskii
Department of Natural Sciences
Federal State Budgetary Educational Institution of Higher Education "Siberian State Industrial University"
Novokuznetsk, Russia

ISSN 2191-530X ISSN 2191-5318 (electronic)
SpringerBriefs in Applied Sciences and Technology
ISBN 978-981-95-1397-0 ISBN 978-981-95-1398-7 (eBook)
https://doi.org/10.1007/978-981-95-1398-7

This Springer imprint is published by the registered company Springer Nature Singapore Pte Ltd.
The registered company address is: 152 Beach Road, #21-01/04 Gateway East, Singapore 189721, Singapore

Preface

Progress in modern engineering and technology is becoming increasingly demanding on mechanical and performance properties of materials. This makes necessary either to employ expensive materials with the required set of properties or modify the structure of existing materials *via* the surface treatment.

In order to protect metals and alloys from different external influences (wear, corrosion, and static and dynamic loads), mining, construction, and metallurgical industries utilize hard-facing steel that provides materials with a set of high mechanical properties. Fabrication of coatings with high-performance properties ensuring higher reliability and longer life time of products in extreme conditions (high mechanical loads, wear, corrosive environments, and thermal cycling) is a foremost problem, which can be solved by the development and implementation of new hard-facing materials containing chemical species, hard alloys, and solid solutions. The coating material is a key determinant of hardening; it differs from the base material and provides the required properties for the surface of parts. High-speed steels are used very often for this purpose. These steels are heat-resistant, i.e., they can retain their properties at high temperatures. Heat resistance is estimated by hot hardness, in accordance to which modern high-speed steels are classified into three categories. The types 1.3355, R12, R9 tungsten and 1.3343, R6M3, R8M3 tungsten–molybdenum steels possess normal heat resistance.[1] The higher heat resistance is typical for 10R8M3 and 10R6M5 steel types with the higher carbon content, R12F3, P2M3F8 steel types with the higher content of vanadium, R18F2K5, R6M5K5, R9K10 steel types with the higher cobalt content, and others. Steels having the high heat resistance usually contain the lower carbon content, but are high in alloying elements, e.g., V11M7K23, V14M7K25, and 3V20K20Kh4F.

To date, high-speed steel (HSS) remains one of the main tool materials. At the global level, the HSS production exceeds 300 thousand tons. The composition of different HSS types contains tungsten 18%, cobalt 10%, cobalt 3–7%, and vanadium 10%. These metals are extremely deficit, and the cost of long HSS is 10 times higher

[1] Hereafter steel grades correspond to Russian standards and have no foreign analogs.

than that of carbon and structural steels. However, in fabricating tools from long steel, the metal utilization factor is not over 0.4.

HS steels are the main material in the production of cutting tools, i.e., their share is ~ 70%, 20% belong to hard alloys, 8% belong to carbon steel, and 4% fall on cemented-oxide and superhard materials. A HSS cutting tool removes almost 80% of the total metal mass from workpieces.

There are carbide-hardened HS steels: carbides precipitated in such steels are responsible for their hardening, providing high mechanical properties. Carbide-hardened low-alloy steels can be divided into three groups. The first group includes ~ 3% W, 3% Mo, 2.5% V, 4% Cr, and 1% C (e.g., the type 11RZAMZF2 steel). Steels from the second group do not contain tungsten; their composition includes ~ 1% C, 2.5–3.5% Mo, 2–3% V, and 4–6% Cr (e.g., the type EP973 steel). Steels from these two groups are microalloyed with nitrogen, niobium, or titanium to reduce their overheating. However, microalloying reduces the steel strength due to the additional formation of nonmetallic inclusions.

The third group includes steels containing 5–6% Mo, 1–2% V, ~ 4% Ti, and 1% C (e.g., the type 11M5F steel) and steels with 1–2% V and tungstenless steels (e.g., R2M5, ROM4, and ROM4F steel types). Performance properties of these steels are higher than that of steels from the first and second groups.

Structural transformations, heat resistance, and tool efficacy are determined by the structural homogeneity. Carbide inhomogeneity is an important property of HSS, which is determined by GOST 19265–73. It should, however, be noted that the issue of carbide inhomogeneity in the structure of cast HSS has not yet been resolved and is not undisputed in terms of the modern theory of the structure and phase composition formation. Some of researchers place high demands on the HSS structure, saying that any manifestation of inhomogeneity considerably reduces its cutting properties. On the other hand, there is a trend which does not so rigidly associate the HSS efficacy with carbide inhomogeneity. The literature illustrates examples of moderate carbide inhomogeneity that in no way proves the defect structure of HS steels.

Along with the volume, uniformity, and principle of the carbide distribution in the steel structure, it is important to mention the real size of the steel grain. During heating for quenching, small grains along carbide grain boundaries melt first that inevitably leads to the austenite grain growth and a decrease in the steel toughness.

In recent years, research and development have been intensively conducted into plasma surfacing (1.3355, R6M5, R2M9) using nitrogen as an alloying element. This alloying is the best, since it allows to considerably improve the corrosion and impact resistance, strength, abrasive wear, and other properties of alloys. This becomes possible due to the improved microhardness of structural components of coatings. Reverse-polarity plasma surfacing with noncurrent carrying cored filler wire in the protective nitrogen medium has a number of advantages over other surfacing techniques. Rationale for the selection of the hard-facing material, meeting the operational conditions, must be proven by extensive studies of its structure, phase composition, mechanical and tribological properties, and their evolution during successive heat treatments.

Microfracture is usually associated with microdefects in surface layers. In the first place, high-quality methods should be developed to form the layer on the workpiece surface. Wear and corrosion resistance can be provided by the surface modification to the wear limit value. The workpiece failure very rarely occurs deep in the bulk material, and thus, the latter requires neither hardening nor phase transitions. This problem can be solved by conventional hardening (thermochemical, mechanical, physical, and chemical) techniques that do not always provide good adhesion between the deposited material and substrate. In order to provide the uniform structure, phase composition, and surface properties at significantly increased functional properties of the material, it is advisable to use electron beam processing. It is one of the promising techniques of the surface modification, its additional hardening, and improvement of mechanical properties of metallic materials. The properties of electron beam-melted metals and alloys can be improved by 20 times that substantially exceeds the efficiency of conventional processing techniques. In a very short time, high-density electron beams induce different changes in the surface layers, namely rapid recrystallization, polishing, and annealing. Moreover, electron beam processing can induce plastic strain in the surface layer, thereby generating high-density dislocations and significantly improving its properties.

The purpose of this work is to investigate the evolution of the structure, phase composition, and properties of the type 1.3355 high-speed steel after tempering and electron beam processing with a view to develop physicotechnical bases of hardening HSS products having different shape and size. This purpose, as well as the identification of improved functional properties of the hard-facing material, can be achieved *via* scanning electron microscopy and up-to-date methods of material physics.

The authors like to express their gratitude to Profs. A. N. Smirnov and V. A. Klimenov for the monograph review and valuable comments, Prof. I. Yu. Litovchenko for his help with scanning electron microscopy observations, Prof. Yurii F. Ivanov for discussions, and A/Professor N. N. Malushin for providing test specimens.

Tomsk, Russia Aleksandr I. Potekaev

Introduction

Development and production of high-speed steel (HSS) began late in the nineteenth century. The Industrial Revolution in the eighteenth century took metalworking production to a new level, although simple hand lathes were used as early as five centuries BC. The performance of hand lathes was slow, low quality, and time-consuming. Mezenin [1] reported that machining of wooden parts, nonferrous metals, and soft steels was performed by well-suited cutters made of hardened carbon steel, but in processing steel parts, these cutters quickly heated, worn, and the part could not be machined at a speed higher than 5 m/min. It was necessary to create steel with good mechanical properties, which at high cutting speeds could withstand high temperatures.

This problem was partially solved in 1858, when Robert Forester Mushet created steel containing 1.85% carbon, 9% tungsten, and 2.5% manganese. Ten years later, Mushet fabricated new steel called self-hardening steel (Mushet steel). It contained 2.15 % carbon, 0.38 % manganese, 5.44 % tungsten, and 0.4 % chromium. In three years, Samuel Osborn started to manufacture R. Mushet's Special Steel in Sheffield. That steel did not lose its operational properties at 300 °C heating and promoted 1.5 times increase in the cutting speed (from 2 to 7.5 m/min). 40 years later, Taylor and Watt, American mechanical engineers put high-speed steel on the market. According to Payson [2], the cutting speed of cutters made of that steel reached 18 m/min. That steel was a prototype of the contemporary steel type 1.3355. And ultrahigh-speed steel was developed 5 or 6 years later; its cutting speed reached 35 m/min. For a 50-year period, the cutting speed increased by 7 times thanks to tungsten, and, consequently, the performance of cutting machines increased by the same amount.

First high-speed steel and methods of engineering testing of cutting tools appeared in 1906. The carbon steel treatment was used as a reference; the cutting speed provided a 20-min tool durability. Tungsten was further used to create cemented carbides consisting of tungsten, chromium, and cobalt. Stellite was fabricated for cutters. The cutting speed of the first created stellite was 45 m/min in the temperature range of 700–750 °C. As reported by Haufe [3], the widia alloy produced by Fried. Krupp AG in 1927, scored of 9.7–9.9 on Mohs hardness scale, and cf. diamond's hardness is 10.

In Russia (the Soviet Union), basic research into high-speed steels was initiated in the years 1930–1935 with the explosion of our industry [4]. HSS grading was specified according to the fact that steel was invented in England, where the high-speed steel was called 'rapid steel'. HSS grades were then identified by the Latin letter *R* denoting 'rapid'. A number after the latter *R* denotes an average content of tungsten in weight percent.

Steels with the low-tungsten content were searched for economic reasons and tungsten deficiency, resulting in over new 20 steel grades. Substantial results were obtained in studying the tool surface hardening. Prokoshkin [5] studied HSS cyanide hardening and determined the nature and properties of the layers obtained. Since 1938, cyanide hardening has been widely used in industry and increased the tool durability by ~ 1.5 times.

Recently, plasma alloying of the HSS surface (plasma surfacing, pulsed plasma alloying, and plasma-ion assisted deposition) has been intensively developed. Different thermochemical methods (cyaniding and boriding) are widely used nowadays. Ultrasonic surface hardening and other ways of the surface alloying are effectively implemented, thereby modifying the surface structure, phase composition, and properties. A combination of various hardening techniques allows generating working layers possessing enhanced performance properties.

The HSS production utilizes a conventional technique of teeming and molding followed by peeing. This technique includes preliminary annealing and quenching of the material, preventing it from brittleness and improving the quality of tools. Another method is metallic powder spraying in nitrogen. Steel manufactured by the first technique is mostly used in the production of cutting tools, whereas steel obtained by the second technique is additionally used as a hard-facing material to impart specific properties to the surface, e.g., cast rolls [6–8]. Such a symbiosis is cost-effective, as plasma surfacing is applied to a less expensive substrate.

However, this production technique of products with a wear resistant, red-hard surface raises a spectrum of questions concerning the formation of structure and phase composition, distribution of hardness and chemical elements in the surface area of the deposited layer, shielding-doping nitrogen atmosphere effect on the structure and properties of the deposited layer, alloying effect on the formation of residual stresses, and prevention of pore, crack, and defect generation in the microstructure during plasma surfacing of heat-resistant steels (62–66 HRC). It is necessary to evaluate the contribution of the controlled thermal cycle of plasma surfacing, hard-facing materials, and high-temperature tempering to the physical nature of hardening of the heat-resistant alloy obtained by plasma deposition in nitrogen.

Electron beam processing is one of the advanced techniques of the HSS surface hardening. Nevertheless, the nature and effect of the electron beam interaction with the steel structure and properties are not always clear. The aim of this monograph is to study these issues.

Table 1 Process conditions for the plasma surfacing system UD-41

Arc current, A	Arc voltage, V	Deposition rate, m/h	Wire feed speed, m/h	Zenith displacement, mm	Arc length, mm
140–160	50–55	18	60	10–12	20

Materials and Methods

Plasma surfacing with noncurrent carrying cored filler wire of a diameter 3.7 mm made of 1.3355 steel grade was used to fabricate test specimens. Plasma surfacing was performed in nitrogen atmosphere onto the ČSN 14331 steel, comprising 0.3 wt.% C, 0.9 wt.% Cr, 0.8 wt.% Mn, and 0.9 wt.% Si. The chemical composition of 1.3355 steel grade consisted of 0.87 wt.% C, 4.41 wt.% Cr, 17.00 wt.% W, 0.10 wt.% Mo, 1.50 wt.% V, 0.35 wt.% Ti, 1.15 wt.% Al, and 0.06 wt.% N. Process conditions included 20–22 L/min shielding gas (nitrogen) rate and 6–8 L/min plasma gas (argon) rate. Table 1 presents process conditions for the plasma surfacing system UD-417 [9–12].

The steel specimens were investigated in three states, namely: after plasma surfacing, after plasma surfacing and 580 °C tempering for 1 hour (the number of cycles 4) [9–12], and after plasma surfacing, high-temperature tempering, and pulsed electron beam irradiation. The SOLO installation (Institute of High Current Electronics SB RAS, Tomsk, Russia) was used for the pulsed electron beam irradiation. The electron beam parameters were as follows: 18 keV energy, 30 J/cm^2 density, 50 μs pulse duration (number of pulses: 5), 0.3 s^{-1} pulse frequency, and 0.02 Pa residual pressure in the chamber.

After electrical discharge machining, all the workpieces were grinded on an abrasive machine with fine sand paper and polished with the diamond paste and, finally, thinned in an electrolyte.

The Olympus GX51 inverted metallurgical microscope was used to investigate the fine structure of the obtained materials. The specimens were etched in a 4% solution of nitric acid and alcohol for 10 s. A KYKY-EM6900 scanning electron microscope (SEM) equipped with an energy dispersive spectroscopy (EDS) detector allowed to analyze the morphology and elementary composition of the modified surface. The KYKY-EM6900 had a thermionic filament made of tungsten. Its operating parameters included 20 kV accelerating voltage, 150 μA emission current, 2.4 A filament current, and 15 mm distance between the specimen and lens [13–16].

Using a digital microVickers hardness tester HVS-1000 equipped with a four-sided diamond pyramid tip, microhardness was measured at 1 N indentation load. A focused ion beam TTX-NHT nanoindentation tester fitted with a Berkovich three-sided diamond pyramid tip was used to study the hardness (nanohardness) properties of deposited coatings at a 50 mN load, 10 Hz acquisition rate, linear loading, 100 mN/min loading and unloading rates, and 5 s interruption.

A pin on disc and oscillating TRIBOtester (Tribotechnic, France) permitted the measurement of Coulomb friction and wear properties of the coating synthesized. Testing conditions included room temperature, 2 mm radius of friction groove, 100 m travel, 10 N load, and Al_2O_3 balls of a diameter 6 mm used as counter bodies.

References

1. N.A. Mezenin, *Interesting About Iron [in Russian]* (Metallurgiya, Moscow, 1972), 200 p
2. P. Payson, *The Metallurgic of Tool Steel* (New York, London, 1968)
3. W. Haufe, *Schneilstahle und ihre Warmebehandlung* (Munchen, 1972)
4. Y.A. Geller, Development of high-speed steels in the past fifty years. Met. Sci. Heat Treat. **9**(11), 810–814 (1967)
5. D.A. Prokoshkin, Diffusion of elements into solid iron: chemical and thermal methods of steel processing [in Russian], in *Trudy Instituta stali* (Metallurgizdat, Moscow, 1939)
6. N.N. Malushin, Integrated use of plasma surfacing and nitriding processes [in Russian], in *Proceeding of 7th All-Russian Science Conference "Innovative Technologies and Economics in Mechanical Engineering"* (TPU, Tomsk, 2009), pp. 58–60
7. N.N. Malushin, D.V. Valuev, Plasma surfacing and nitriding of plasma surfaced parts of mining and metals production sector. Min. Inf. Anal. Bull. (Sci. Tech. J.) **12**, 105–108 (2012)
8. A.P. Kovalev, N.N. Malushin, D.A. Romanov, Heat treatment effect on structure of coatings formed by plasma surfacing in nitrogen atmosphere with chromium-tungsten metal of P18 steel type [in Russian], in *Coll. Papers in memory of I. A. Oding "Mechanical Properties of Modern Structural Materials"* (Moscow, 2020), pp. 49–50
9. N.N. Malushin, D.V. Valuev, *Quality Assurance of Metallurgical Equipment Parts During Life Cycle Using Plasma Deposition of Heat-Resistant Steels [in Russian]* (Scientific Technology Publishing House, Tomsk, 2013), 358 p
10. N.N. Malushin, V.E. Gromov, D.A. Romanov, L.P. Bashchenko, O.A. Peregudov, *Plasma-Assisted Surface Hardening in Nitrogen Environment of Heat-Resistant Alloys [in Russian].* (Poligrafist, Novokuznetsk, 2022)
11. N.N. Malushin, D.V. Valuev, V.L. Osetkovskii, S.A. Solodskii, *Part Surfacing Methods in Mining and Metals Production Sector Using Heat-Resistant High-Hardness Steels [in Russian]* (TPU, Tomsk, 2015)
12. N.N. Malushin, V.E. Gromov, D.A. Romanov, L.P. Bashchenko, A.P. Kovalev, O.A. Peregudov, Sparingly alloyed cored wires and their multilayer plasma deposition in nitrogen environment. Zagotovitel'nye proizvodstva v mashinostroenii **21**(2), 58–63 (2023)
13. M. Ladd, R. Palmer, *Structure Determination by X-ray Crystallography* (Kluwer Academic, Plenum Publishers, New York, 2003)
14. Y. Waseda, E. Matsubara, K. Shinoda, *Diffraction Crystallography. Introduction, Examples and Solved Problems* (Springer, Berlin, 2011)
15. F. Ernst, M. Ruhle, *Science C. High-Resolution Imaging and Spectrometry of Materials* (Springer, Berlin, 2003)
16. T. Weirich, *Electron Crystallography* (Springer, Berlin, 2006)

Contents

Chapter 1
Current Studies of Structure, Phase Composition, and Properties of High-Speed Steels and Their Application

1.1 Modern High-Speed Steels and Fields of Application

High-speed steels are a subset of tool steels currently used in different industries. Their microstructure is characterized by a significant (50 vol.%) carbide amount distributed in the matrix of tempered martensite [1]. This microstructure provides HSS with remarkable surface hardness, resistance to wear, and fatigue cracking both at room and elevated temperatures [2]. Therefore, HSS is widely used to manufacture wear-resistant parts, such as cutting tools, milling cutter, and roll bearings, as illustrated in Fig. 1.1a [3–5]. According to a search in the bibliometric database Web of Science, the number of papers and patents concerning HSS is being constantly increased from 9.1% in 2014 to 11.9% in 2020 (Fig. 1.1b). According to Fig. 1.1c, the global production capacity of this steel increases from 37.7×10^4 tons in 2014 up to 46.8×10^4 tons in 2020. This indicates to a growing interest of researchers and market demand for HSS both in science and industry.

The HSS production includes conventional casting [6–8], electroslag remelting [9–12], and powder metallurgy [13–16]. According to Mesquita and Barbosa [17], in conventional casting, segregation of large amount of carbon and alloying elements before crystallization leads to the formation of an interganular carbide network owing to a slow solidification rate. Post-forging or rolling is therefore required for the carbide network fragmentation that leads to a low output [18, 19]. Electroslag remelting provides relatively high cooling rate, while a large amount of ledeburite together with carbide networks are still being generated, thereby degrading engineering properties of steels [20, 21]. The powder metallurgy process demonstrates advantages of HSS with dispersed and small primary carbides and high mechanical properties. However, this process is limited by its high cost and low performance owing to complicated heat treatment and requirements for precision equipment [13, 22].

Edgar and Tint [23] report that additive manufacturing (AM) allows fabricating mechanical components of a complex geometry, wide property control, and specific

A. I. Potekaev et al., *High-Speed Steel After Tempering and Electron Beam Processing*, SpringerBriefs in Applied Sciences and Technology,
https://doi.org/10.1007/978-981-95-1398-7_1

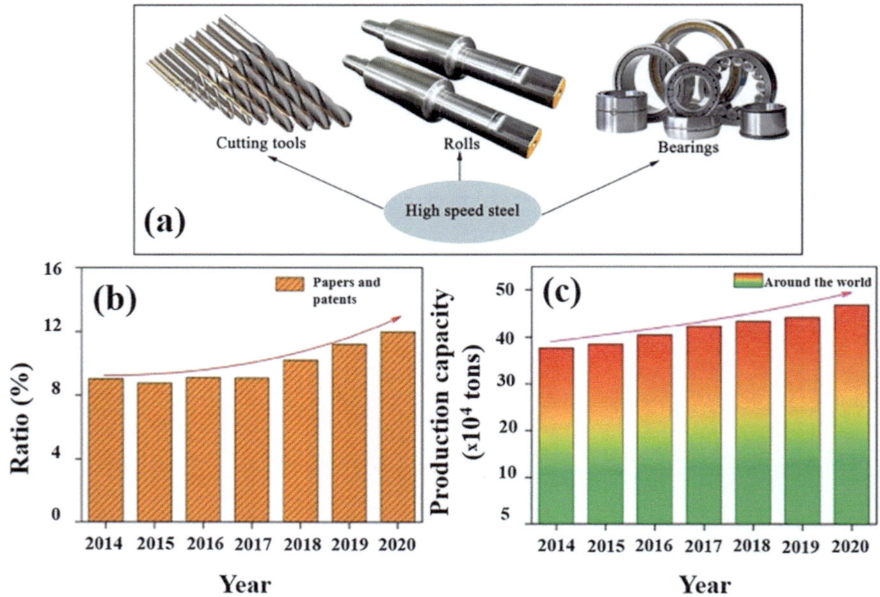

Fig. 1.1 HSS applications (**a**), papers and patents concerning HSS in research community (**b**), global production capacity from 2014 till 2020 (**c**)

quality beyond traditional production techniques. AM methods such as metal powder additive manufacturing, directed-energy deposition, material jetting are investigated as a paradigm to manufacture products for a wide use in automobile, defense, and aerospace engineering [24]. AM of high-speed steels can be an effective alternative to repair of damaged tools, when a damaged surface is coated with a 3D-printed layer. AM methods are currently used in industry to create high-performing and high-speed products [25–27]. Moreover, a rapid development of AM methods considerably expands the production route and spectrum of alloys made of different HSS grades.

In recent years, extensive research has been done into additive manufacturing of HSS. Saewe et al. [28–30], for example, conducted feasibility investigation for the production of M50 and HS6-5-3-8 steel grades, using selective laser meeting (SLM). They showed that in using the appropriate process parameters, dense components without cracks could be fabricated from hardened high-speed steel. Further research [31] was focused on the influence of increased carbon content on processability of high-speed steel HS6-5-3-8 by SLM. It was found that the increased carbon content in the alloy resulted in a completely austenite matrix and a larger carbide content. As opposed to the expected effect of the austenite stabilization on the crack propagation, it intensified with increasing carbon content. Jin et al. [32] described the solidification microstructure and carbide precipitation from the electron beam-melted HSS S390-type high in carbon and alloying elements. Varying the scan rate, Rahman et al. [33] used laser metal deposition of vanadium-rich high-speed steel in order to study the influence of the high cooling rate on carbide distribution and morphology. It was found that the deposition rate improved the microhardness of vanadium-rich HSS. Similar results were obtained by Arias et al. [34, 35] for the AISI M2 high-speed tool steels after laser surface melting. In particular, a multilayer spray forming of steels M4 and Fe–1.4C–5W–4Mo–3V–4Cr (wt.%) was studied by Shim et al. [36] and Rahmana et al. [37]. The thickness of these steels ranged from moderately thin to rather high. Moreover, Lee et al. [38, 39] studied carbide M_2C decomposition and HSS distribution high in cobalt generated by spray forming. It was shown that changes in temperature conditions induced the formation of new carbide morphologies and their distribution during spray forming. Lu et al. [40] fabricated M3:2 high-speed steel by spray forming and niobium alloying that increased thermal stability of carbides M_2C.

Let us mention some of review articles focusing on additive manufacturing of steels. Fayazfar et al. [41] analyzed powder-based additive manufacturing of ferrous alloys and the influence of the most important process on their microstructure and mechanical properties (hardness, tension, and cyclic/fatigue behavior). Bajaj et al. [42] generalized the microstructure evolution of different steel grades and detected the relation between their microstructure and operational properties. Haghdadi et al. [43] focused mostly on mechanical, corrosion, and wear-resistant properties of steels and briefly discussed their microstructure evolution in order to explain the difference in the properties obtained. A review given by Yin et al. [44] for laser additive manufacturing of austenitic, ferritic, and martensitic steels, emphasized the understanding of relations between processes, microstructure, and mechanical properties. Although these review articles highlighted the microstructure and performance specifications of AM-fabricated stainless and martensitic steels, a complete review of different aspects of HSS additive manufacturing is still absent.

With regard to a rapid progress of research in and practical applications of HSS additive manufacturing, we consider our review, and generalization of the latest data will be very timely and meaningful.

AM engineering solutions based on different approaches are considered and compared below. Numerical simulation results are generalized, and different AM processes are analyzed, including the influence of the AM process variables on the HSS microstructure components. Discussed are the mechanical and tribological properties of AM-fabricated HSS such as hardness, strength, fracture toughness, and wear resistance of pars. In addition, the persistent challenges and future research directions in the field of HSS additive manufacturing are discussed herein.

In accordance with ISO/ASTM 52900:2015, there are 7 categories of AM methods, while the literature presents detailed description of many AM methods [45, 46]. Each AM method is intended for the production of a certain category of materials due to their specific formation and deposition techniques. Among all commercially available methods, additive manufacturing of HSS parts is successfully used by various companies. Classification of AM methods is presented in Fig. 1.2 [47].

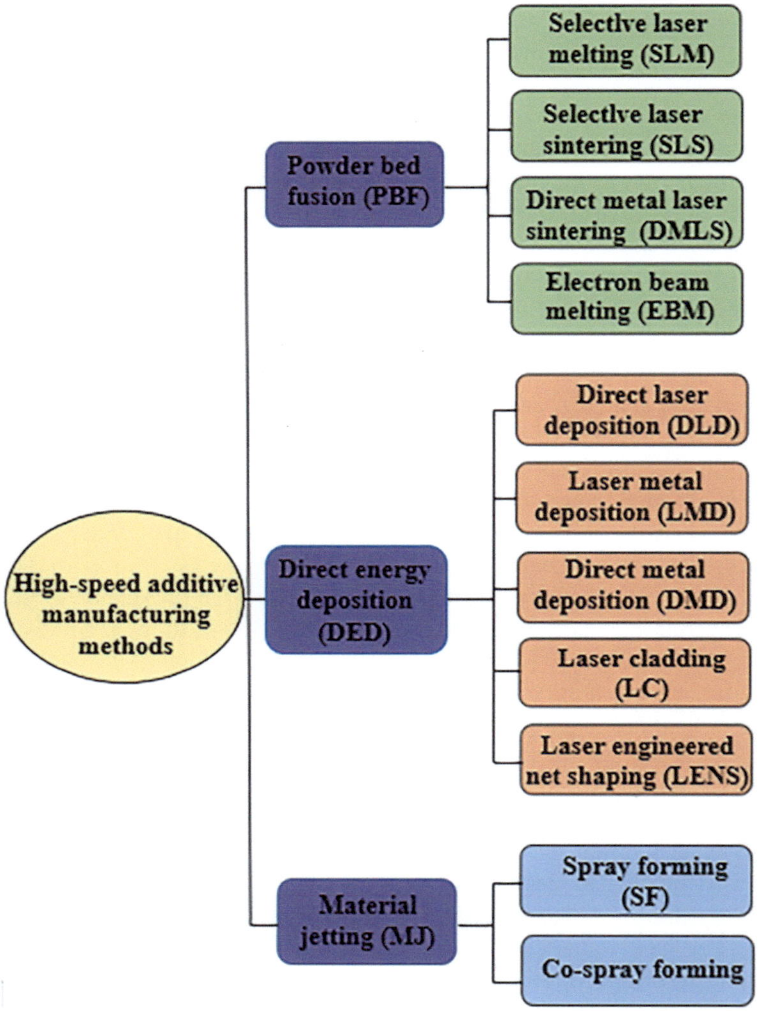

Fig. 1.2 Block diagram of common AM methods of HSS fabrication

1.2 Structure and Properties of HSS After Alloying

Special requirements are placed on high-speed tool steels, namely hardness, toughness, wear and heat resistance, hardness penetration, oxidation resistance, fire cracking resistance, and others. According to Houdremont [48], the properties of such steels are determined by the structure and alloying elements forming this structure. Hardness is one of the main service properties of high-speed steels, which is determined by their chemical composition, primarily carbon, i.e., its volume fraction, carbide phase hardness, and heat treatment [49, 50]. The latter is used to obtain the appropriate steel structure, first of all, the formation of martensite during quenching. The martensite hardness in quenched steel is determined by the carbon content and

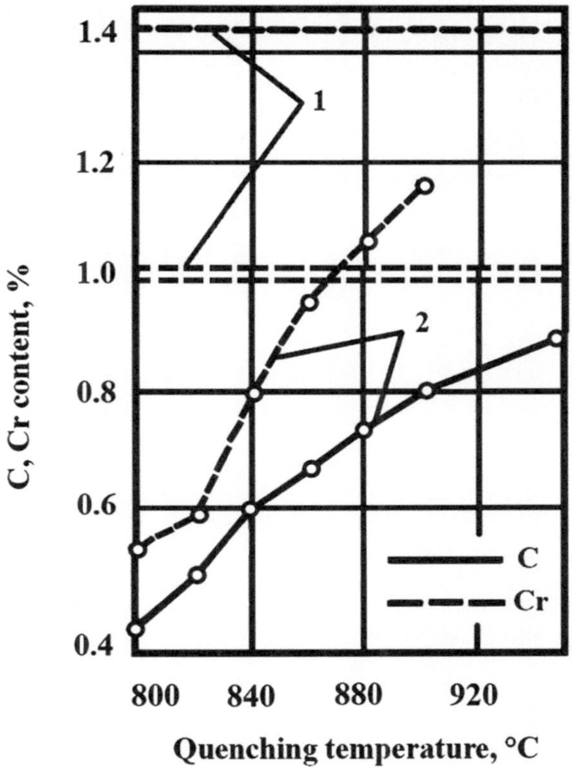

Fig. 1.3 C and Cr content in martensite of hardened steel (1.02% C, 1.4% Cr, 0.43% Mn), depending on quenching temperature: 1—in steel, 2—in a quencher

reduces from 60–62 HRC at 0.6–0.7% C to 45–50 HRC at 0.2–0.5% C and to 30–40 HRC at < 0.1% C. The steel hardness can be slightly higher due to the presence of strengthening phases (see Fig. 1.3).

The HSS toughness is primarily determined by the hardness in addition to the number of nonmetallic inclusions, impurities, and carbide inhomogeneity (for carbide-strengthened steels). Hardness and toughness are antagonists. Most often, the improved hardness leads to the toughness reduction.

Wear resistance of HSS characterizes the operational life of the tool. It is determined by the steel composition, structure, and properties and also the properties of a workpiece. Wear resistance depends, in the first place, on the steel hardness and structure, i.e., the content and grain size of carbide phase, the content of martensite, carbon content, and others. In Fig. 1.4, one can see dependences between the HSS wear resistance and hardness [52].

The hardness growth reduces the tool wear, although it slightly increases at > 64 HRC due to brittle chipping of the cutting edge. In comparison with steel containing 12% W, the growth in the carbide phase in steel with 18% W also leads to a higher wear resistance (Fig. 1.5) [52].

Hot or red hardness is one of the main properties of the HHS determined by the temperature to which the values of hardness, strength, and wear resistance remain

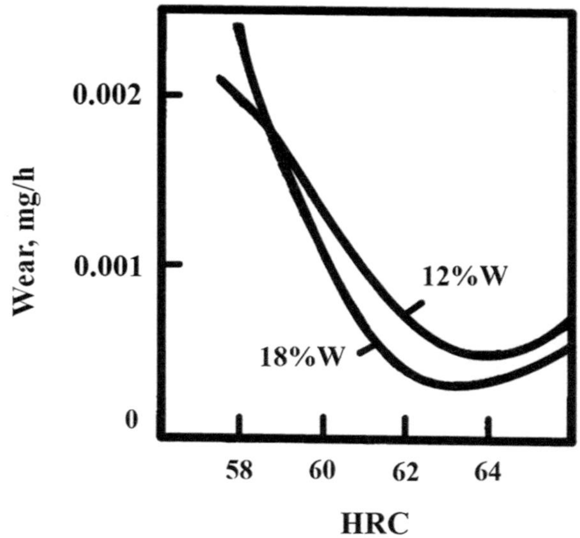

Fig. 1.4 Dependences between the HSS wear resistance and hardness at different W content

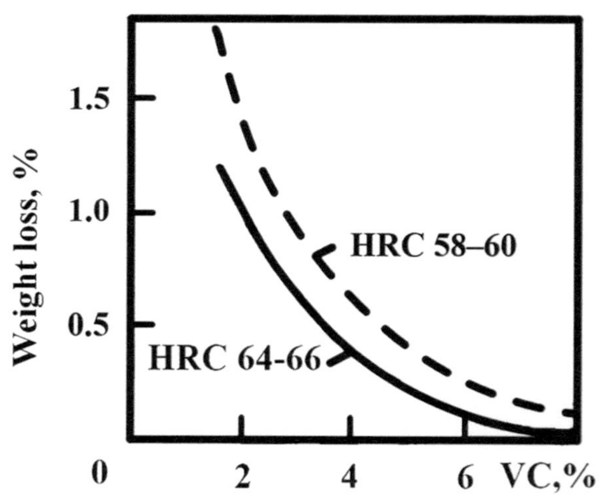

Fig. 1.5 Dependences between the HSS wear resistance and vanadium carbide (VC) content at different hardness

as specified. It means that hot hardness describes the ability of steel to withstand modifications of the structure and properties of the cutting edge during heating and the steel resistance to tempering. High-speed steels offer excellent performance up to 550–700 °C.

Hardness measurements conducted during steel heating (hot hardness) and after cooling (cold hardness) show this different hardness, depending on the heating temperature [53]. There are two temperature regions widely varying during changes in hot and cold hardness. The first region corresponds to 550–600 °C heating, depending on the steel composition. Cold hardness after heating remains unchanged as compared to its initial values that indicates to the fact that the steel structure does not modify within these temperatures. Hot hardness at the same heating temperatures reduces insignificantly and almost proportionally to the temperature growth. Within 200–550 °C, hot hardness decreases due to the heat-induced softening, which is a reversible process.

The second region corresponds to heating higher than 600 °C and is characterized by a drastic decrease (by 300–400 HV for each 100 °C) in hot hardness and cold hardness proportionally to the temperature growth. The hardness reduction is mostly induced by the development of irreversible processes modifying the steel structure, i.e., solid solution (martensite) decomposition and coagulation of disperse strengthening phases generated during prior tempering. When approaching more closely to temperatures of phase transformation, a significant softening of the base metal is observed.

Heat resistance is measured by the heating temperature, at which the structure and properties start to irreversible modify, resulting in the significant impairment of the tool life. It is advisable to measure heat resistance by cold hardness. Such measurements are easy while at the same time identify the steel composition and heat treatment.

Figure 1.6 presents the temperature dependence of hardness for R18 steel grade [53]. This steel withstands temperature up to which its hardness remains 60 HRC.

HSS hardenability describes such properties as, for example, hardness in the cross-section. It is detected by the resistance of undercooled austenite to pearlitic transformation. The steel capability to surface hardening (the maximum hardness on the tool surface) largely depends on its hardenability also.

Oxidation resistance characterizes the steel resistance to gas corrosion. Basics of HS steel alloying are presented in works of [48, 51, 54–56].

In the case of HS steel hardening with carbide precipitation from alloying elements, it acquires high mechanical properties. R18 and R6M5 steels relate to this type. Kremnev [57] classifies low-alloy HS steels as carbide-hardened steels,

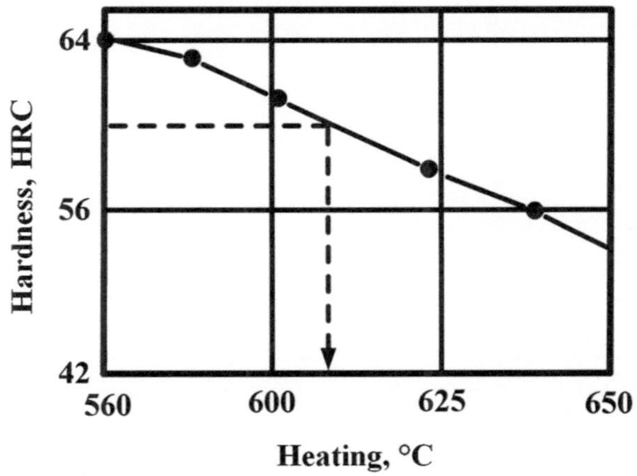

Fig. 1.6 Temperature dependence of R18 steel hot hardness

which can be divided into three groups. The first group consists of ~ 3% W, 3% Mo, 2.5% V, 4% Cr, and 1% C (e.g., 11RZAMZF2). The second one contains no tungsten and includes ~ 1% C, 2.5–3.5% Mo, 2–3% V, and 4–6% Cr (e.g., EP973). In order to reduce a tendency to overheating, steels of these two groups are microalloyed with nitrogen, niobium, or titanium. This, however, decreases their strength due to the additional formation of nonmetallic inclusions. Low processing properties and tendency to overheating restrict the use of these steels instead of R18 and R6M5 grades. The third steel group includes 5–6% Mo, 1–2% V, ~ 4% Ti, and 1% C (e.g., 11M5F) and steels with 1–2% V and tungstenless steels (R2M5, ROM4, and ROM4F). These steel groups are characterized by higher performance properties than steels from the first two groups.

Dispersion hardening of HSS can be obtained by intermetallic precipitation due to the formation of intermetallic phases. Intermetallic-precipitation-strengthened steels include carbon-free and low carbon steels of the iron–cobalt–tungsten–molybdenum system [58], which can be additionally alloyed with nickel, chromium, and titanium. The chemical composition of such steels is presented in Table 1.1 [57, 59]. The chemical composition, physical, mechanical and processing properties, and application fields of these steels differ from those of carbide-precipitation-hardened steels.

In recent years, new high-speed steel grades have been developed and implemented because of the tungsten high cost and lack. These steels (R18F4K8M, R18FZK8M, R18F2K8M, R9F4K8M, and R6F2K15M5) consist of high and medium content of tungsten and carbon increased by 0.2%; their maximum secondary hardness is 68–70 HRC, and hot hardness ranges between 640 and 650 °C. Other grades include high-carbon, low-tungsten, and tungstenless steels (R6F2M5, R6F2K8M5, R6F2K8M6, R2F2K8M6, ROF2K8M6) possessing 63–65 HRC secondary hardness, 620–635 °C hot hardness, and improved processing properties [59].

Chaus and Latyshev [60] investigated the structure and phase composition of as-cast R6M5 steel alloyed with vanadium, titanium, and niobium using the metallographic analysis, X-ray diffraction analysis, and X-ray microanalysis after successive heat treatments. The latter induced a diffusion redistribution of alloying elements

Table 1.1 Chemical composition of intermetallic-precipitation-strengthened steels

Steel grades	Chemical composition, %						
	C	W	Mo	Ni	Cr	Co	Ti
High heat resistance							
V11M7K23 (EP831)	0.10	1.00	7.50	–	–	23.00	0.10
V4M12K23	0.10	1.00	12.50	–	–	23.00	0.10
Increased heat resistance							
I8K14M18T	0.03	4.00	17.50	–	–	13.50	0.60
Low heat resistance							
03M10Kh11M2T2	0.03	–	2.00	10.00	11.00	–	2.00

between the base metal and excess phases, while the carbide components underwent qualitative and quantitative changes. The alloying process significantly modified the base metal structure and carbide phase of the high-speed steel.

Quite recently, considerable attention has been paid to the development of new alloying systems and improvement of HSS performance properties, especially in China, where economies experience a fast growth. Jiao et al. [61] studied the advanced AISI M42 high-speed steel with the high (0.18 wt.%) nitrogen content, fabricated by metal injection molding. High-temperature softening resistance of this steel without N and the type 0.18 N steel was evaluated by hardness measurements at room temperature after additional heating at 600 °C for different time. Wear resistance at 600 °C was examined using the ball-on-disc method. SEM observations, X-ray diffraction (XRD) analysis, and high-resolution transmission electron microscopy (TEM) were used for the microstructure characterization. It was found that the hardness of the 0.18 N steel was 2–3 HRC higher than that of the steel without *N* after the additional heating at 600 °C for 4–120 h. The excellent high-temperature softening resistance of the 0.18 N steel was explained by the lower growth rate of secondary carbides M_2C. In ball-on-disc tribology testing at 600 °C, the wear rate for the steel without *N* was ~ 1.5 times higher than for the 0.18 N steel. The best high-temperature softening resistance of the 0.18 N steel was mostly attributed to the fact that its excellent heat resistance allowed it to retain high hardness during wear, substantially strengthening the oxide layer and increasing the steel protection.

Li et al. [62] studied the effect of magnesium treatment on inclusions, microstructure, and mechanical properties of M7 high-speed steel under pressure. It was shown that magnesium can effectively increase the steel purity, as the content of oxides, *S* and *N* gradually reduced with increasing Mg content. And if the content of coarse inclusions substantially reduced, then the content of fine inclusions gradually increased. Moreover, the type of inclusions was changed according to simultaneous $Al_2O_3 \rightarrow MgO \cdot Al_2O_3 \rightarrow MgO$, $MnS \rightarrow MgS$ transformations. Magnesium treatment significantly improved the size of grains and dendrites. The network structure of carbides was broken, and the segregation process clearly weakened. The steel plasticity and fracture toughness were enhanced.

Microalloying used in [63–65], also modified the structure and performance properties of HSS. It was demonstrated that microalloying combined with heat treatment of HSS workpieces enhanced wear resistance of the cutting edge. The modified steel structure after such a combined treatment resulted in changes in its hardness and microhardness. When working with workpieces and in a production environment, microalloying combined with heat treatment led to an increase in the cutting tool life.

Interesting is the tool production based on the steel powder sintering. Akhmetov and Eremeeva [66] investigated the microstructure of a hot-pressed R6M5K5 workpiece consisting of 6 wt.% W, 5 wt.% Mo, 5 wt.% Co, 0.9 wt.% C, 4 wt.% Cr, 2 wt.% V, and Fe residue and diffusion-alloyed additive based on tungsten, molybdenum, cobalt, and iron. The R6M5K5 workpiece was subjected to oil quenching at 1230 °C and double high-temperature tempering at 550 °C for 60 min. The microstructure was examined on a SEM with an EDS analyzer. After quenching, fine-grained carbides dissolved in the matrix, whereas after double high-temperature tempering, precipitation of secondary carbides (M_6C and MC) occurred. Secondary carbides appeared from a separate structural component, i.e., ferrovanadium phase, the powder of which was used as a vanadium master alloy. A nonuniform distribution was observed for cobalt locating in a separate phase based on the diffusion-alloyed additive. These phases did not dissolve and were rather stable in hot-pressing conditions and heat treatment. It was shown that the workpiece hardness changed from 63.1 HRA after hot pressing to 67.2 HRA after quenching and to 66.3 HRA after double tempering.

Although the classical HSS grades R18, R6M5 and low-tungsten grade R9 have been well investigated, and their structure and properties are still being studied [67]. High-performance properties of HSS tools are gained via heat treatment. Therefore, workpieces were cut from the cutting tool made of R6M5, R9, and R18 steels subjected to a conventional treatment. It was found that the initial structure of these steel grades, i.e., after heat treatment, consisted of martensite and special carbides. Carbide particles uniformly distributed in the matrix had almost spherical shape. The structure of R6M5 and R9 steels consisted of M_6C, MC carbides, whereas the R18 steel structure consisted of M_6C carbide only. According to the electron backscatter diffraction (EBSD) analysis, M_6C carbide fitted best the cubic Fe_3W_3C phase, and MS carbide fitted the VC phase. The SEM analysis showed that in addition to M_6C and MC carbides, the HSS structure contained traces of cemented carbides.

The effect of niobium on the secondary precipitates and tempering resistance of spray-formed M3:2 high-speed steel was studied in [68]. Spray forming with the addition of 2% Nb and without it was conducted at 600 °C double tempering. Carbides were clearly identified by a TEM. The steel hardening in the most tempered state occurred due to coherent precipitation of M_6C and MC carbides. After repeated tempering, the M3:2 steel hardness significantly reduced due to M_2C coarsening with the M_3C precipitation, while precipitation of fine and dense M_2C carbide was responsible for the unique long-term strength of Nb-containing steel. MC carbides demonstrated a normal stability. The Nb addition to the substantial Mo and Cr dissolution in the matrix prior to quenching, enhanced the dispersion of secondary M_2C carbide and retarded carbide coarsening that resulted in an excellent tempering resistance.

Although the classical steel R18 contains ~ 1% V, modification of spray-formed HSS with vanadium is nonetheless interesting [69]. Spray-formed vanadium-modified high-speed steel was subjected to various heat treatments, and the obtained microstructure and mechanical properties were investigated. High-temperature Vickers indentation associated with elevated temperature tension test was used for hardness and residual stress measurements of annealed specimens. In addition, studied was the quenching temperature effect on the steel hardness, flexural strength, and fracture toughness. The behavior of secondary carbides and its effect on the steel hardening correlated with TEM observations. A research method was proposed for the behavior of carbide precipitation during heat treatments. It was suggested that our research findings could help to understand the evolution of microstructure and mechanical properties of spray-formed HSS after thermal treatments and optimize the alloy design and post-thermal mechanical treatment.

Ouyang et al. [70] analyzed the CeB_6 effect on the microstructure and mechanical properties of HSS consolidated by powder metallurgy. The M3:2 steel was manufactured by sintering with the addition of different CeB_6 content. The latter improved the density of the sintered steel and provided the formation of the M_6C phase in the microstructure. It was expected that CeB_6 decomposed into boron and cerium. B atoms enriched the M_6C phase, while Ce atoms located mostly at the grain boundaries when transforming to oxides, thereby enhancing the mechanical properties. For example, 1210 °C sintering resulted in ~ 98.5% relative density and ~ 18 μm average grain size at 0.3 wt.% CeB_6. The steel specimen comprising 0.3 wt.% CeB_6 demonstrated excellent mechanical properties, viz. ~ 52 HRC hardness, ~ 3.05 GPa flexural strength, and ~ 40.92 MPa · $m^{1/2}$ fracture toughness, i.e., these parameters increased by 23.5, 38.1, and 23.7% as compared to a specimen without the CeB_6 addition.

1.3 HSS Hardening Methods

The HSS surface hardening can be achieved by different widespread methods, including conventional thermochemical processing [71–76], physical and chemical vapor deposition [77–80], and laser cladding [81–83]. A review of these methods is presented in [84].

In the last few years, there has been a development of new methods of the surface modification and strengthening of HSS products. Plasma-assisted deposition of thin films is the most important, as it endows the product with new surface properties at normal temperature while not changing its dimensions. This expands the performance properties of materials in addition to those gained by the classical thermal treatment. Thin film deposition makes the surface operate as a composite material, the properties of which differ from the properties of both base metal and coating material that hampers the prediction of the product workability as a whole [85].

Plasma-assisted deposition of 1-μm thick coating of the Si–O–C–N system onto the R6M5 steel surface improves its resistance to mechanical damage in addition to a conventional heat treatment. This coating is inherently brittle, and at high contact pressures, the positive effect is primarily provided by the hardened near-surface layer of the base material, since the coating spalls during the initial contact. At medium and low contact pressures, the positive effect is achieved by the strength of the coating and near-surface layer of the base material as well as the reduced surface roughness. The Si–O–C–N system hardening is desirable for products operating at normal contact pressures.

In mechanical engineering, ion nitriding is one of the effective thermochemical processing methods. This is because ion nitriding improves mechanical and performance properties of parts and tools, namely strength, wear resistance, crack resistance, corrosion resistance, fatigue life, and thermal stability. Ion nitriding has, however, a serious limitation: the low diffusion rate. Mamontov et al. [86] report that this limitation can be eliminated by severe plastic deformation after nitriding in glow discharge plasma. The specimen surface after torsional strain demonstrates a refined grain structure [87], which leads to an increase in the surface free energy [88]. In turn, this leads to a higher adsorption of the saturating element and nitride formation in the near-surface layer and provides the higher surface hardness. Due to the increased dislocation density, microdefect formation, and grain refinement, the diffusion rate of the saturating element grows [89, 90], promoting the thickness growth of the hardening layer [86].

Preliminary severe plastic deformation significantly affects the depth and growth rate of the strengthened layer at nitriding for 2–4 h. The maximum depth gain is 60 μm (120%) during 4 h nitriding, while 6 h nitriding shows no effect from preliminary severe plastic deformation [86].

It is also interesting that the depth of the strengthened layer after 6 h nitriding is lower than the depth obtained after 4 h nitriding. Therefore, the steel specimen subjected to severe plastic deformation softens after 6 h nitriding. This issue requires further investigation [86].

Valiev and Aleksandrov [87] focused on a promising surface treatment combining ion nitriding and (Ti, Cr)N coating deposition.

Migranov et al. [88] reported that tribological properties of multilayer coatings for cutting tools were improved by a preliminary two-stage hardening of the surface layer, i.e., the addition of triboactive elements into nitride hard coatings followed by a deposition of wear-resistant coating (Ti, Cr)N. Such multilayer coatings increased the tool life by 2.1–2.4 times due to a longer stage of normal wear. The best combination of high strength and reliability, described by the high coating adhesion to the substrate, was manifested by the multilayer coating with the bottom layer enriched with indium (In) presenting in both free and bound (In–N) states. Wear resistance of HS steel cutting tool with the coated surface was studied for 16 chemical elements and 4 antifrictional materials implanted in the base surface. It was shown that coatings with the modified bottom layer with high antifrictional properties had the highest wear resistance. A positive indium effect on the wear resistance was probably associated with two phenomena observed on the friction surface. Acting as a liquid lubricant, this soft metal reduced the friction coefficient. When In was heated during cutting, it partially oxidized, resulting in the formation of oxygen-containing phases that protected the surface from catastrophic wear.

In [89], the coating was generated in two stages, namely nitriding and cathodic arc plasma deposition of the (Ti, Cr)N coating. The proposed coating also included the additional ion-implanted layer deposited onto a preliminary nitrided surface of the high-speed steel. Such a multilayer coating increased the wear resistance of cutting tools by 2.1–2.4 times due to the prolonged stage of normal wear.

Laser surface hardening is being increasingly used to obtain the desired physical and mechanical properties, structure, and phase composition [90]. In works [91, 92], physicotechnical bases are developed for laser hardening of the HSS surface. It is found that the initial hardness 63–65 HRC of hardened R6M5 steel reduces by 1 or 2 units HRC at the following laser beam parameters: 2 mm beam diameter (0.0314 cm^2 beam area), 0.8 cm/s travel speed, and 12–16 kW/cm^2 power density (376.8–502.4 W power). The hardness reduction in the heat-affected zone (HAZ) is induced by partial dissolution of tungsten carbides. The martensite structure does not change [90].

The development of laser surface hardening, in particular up-to-date high-power laser systems, promotes the development of new heat treatment methods of metals. There is, however, no information in the literature about their implementation. Laser surface hardening of tempered HS steels still remains a relevant problem.

Kupalova [93] compared HSS hardening and fine structure and properties in the initial state and after the additional alloying with cobalt, thermomechanical treatment, and stepwise steel tempering.

Thermal treatment of HSS with use of reduced stepwise tempering method differing from 1 h 560 °C tempering demonstrates a significant improvement of mechanical and performance properties without the additional expensive alloying or complex process operations. Favorable structure and phase composition (e.g., a higher dispersion of carbides) of steel can be obtained by a reasonable selection of the steel grade and the best tempering conditions, including stepwise (reduced) tempering. Changes in the carbide structure and phase composition after heat treatment of high-speed tungsten–molybdenum steel are analyzed in [94]. The nature of carbides precipitating during the eutectic reaction depends on both HSS chemical composition and melt solidification conditions, primarily the cooling rate [95, 96]. The most widespread R6M5 steel refers to tungsten–molybdenum steels. The eutectics formation is based on M_2C, M_6C, and MC carbides, whose volume fraction can be changed in the cast steel. The eutectics content in the alloy can be controlled by inoculation of a small content of pure elements or their carbides, nitrides, and borides and also by varying the rate of solidification [97].

The presence of coarse carbide particles of the eutectic nature is observed in the microstructure of tempered steel. These carbide particles have a two-phase structure, i.e., the chemical composition at the particle center matches M_2C carbide, while the structure on its periphery with MC formations matches M_6C carbide. Moreover, there are small rounded particles incompletely dissolved during austenitization, in which the vanadium content corresponds to M_2C carbide [94]. All this proves incomplete decomposition of metastable M_2C carbide during heating at hot plastic strain and successive thermal processing of HSS that has not yet been reported in the literature (Fig. 1.7).

Aerodynamic sound hardening [98] is a promising method of the HSS treatment. This method utilizes low-temperature thermal energy produced by acoustic resonance affecting the R6M5 steel and (i) eliminates the structural segregation banding and carbide phases along the strain axis due to a reduction in the structural anisotropy and phase hardening conditioned by the stress relaxation and dislocation density; (ii) decreases carbide inhomogeneity by 1 point; (iii) reduces the size of primary carbides from 7–15 to 3–8 μm and their content by 1.6 times. Thus, aerodynamic sound hardening significantly improves the R6M5 steel structure.

Tokarev et al. [99] investigated the high-power laser effect on the hardness and microstructure of the hardened R6M5 steel in order to provide heat treatment using laser irradiation. Hardness measurements in the HAZ showed that the quenched steel hardness decreased by 1–2 units HRC that allowed to conclude about the steel temper.

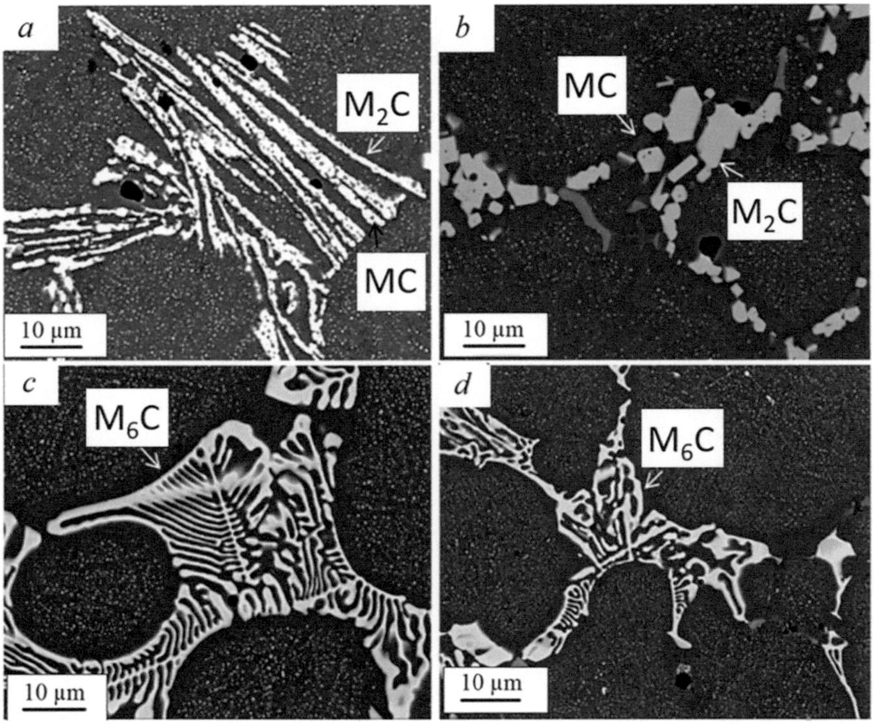

Fig. 1.7 SEM images of R6M5 cast steel microstructure after quenching and tempering [94]: **a**, **b**—decomposition products of M_2C carbide, **c**, **d**—M_6C morphology; **a**, **c**—1200 °C austenitization, **b**, **d**—1240 °C austenitization

Microstructure measurements showed that the HAZ ablation was caused by a partial dissolution of tungsten carbides, although the martensite structure did not change. It was thus shown that the quenched R6M5 steel could be tempered on an Optomec LENS 850-R is 3D printer using high-power laser.

In the most of literature, metal bond diamond grinding wheels are recommended for the HSS polishing based on electrophysical, electrochemical, or combined treatment techniques [100]. At the same time, it is advisable either to restrict the scope of application by the circular grinding by virtue of the defect layer formation on the treated surface or minimize processing parameters. The defect layer is manifested by repetitive cleavages along the cutting edge and microcleavages, size of carbide particles, craters or dimples on the front surface of specimens, steel softening, and higher roughness height.

Lavrent'ev and Dozhdelev [101] studied the HAZ structure and properties of the type 30HGSA (ČSN 14331) steel, forming during DC inert gas arc welding of powder wire made of the R2M8 HSS in fabricating bimetallic tool. It was found that the largest decrease in the steel hardness occurred in the overheated and complete recrystallization area. The steel hardness reduction was also observed in the area of high-temperature tempering. It was demonstrated that the steel hardness could be enhanced due to an increased cooling rate in the HAZ in minimizing the heat contribution to the arc welding process using dual control for metal transfer. Experiments demonstrated the higher steel hardness in the HAZ gained by thermal cycle surfacing control during the bimetallic tool fabrication.

Cherenda et al. [102] investigated the phase and elemental compositions and the surface microhardness of the NbC–HSS system subjected to a combined action of compression plasma flows and heat treatment. The structure and properties of the modified layer were studied by the XRD analysis, SEM, EDS analysis, and microhardness measurements. These studies showed that plasma processing of the NbC–HSS system induced the carbide dissolution and formation of niobium nitride films ε-NbN and β-Nb_2N. The Nb content in the layer reduced with increasing energy density absorbed by the surface. The carbide dissolution led to the microhardness decrease. Compared to the initial steel, the microhardness increased by 1.3 times after successive annealing at 600 °C for 9 h.

Let us consider investigations conducted abroad. Combined thermochemical processes of the surface alloying and nitriding turn to be efficient in surface hardening of different steels [103]. This work focuses on a combined saturation of the R6M5 steel with W and N_2. The layer structure includes the inner nitriding zone comprising the solid solution of tungsten and nitrogen in iron and dispersed inclusions of tungsten nitrides. Dispersion and solid solution strengthening provides a double growth in the microhardness of the modified W–N layer as compared to the base metal. A transition diffusion zone of nitrous martensite formed under the quenched layer creates the main microhardness gradient from the layer to the center and prevents it from embrittlement, peeling, and pigmentation. Based on the metallographic analysis, the hardening rate (yield strength growth) of the modified layer is calculated by the proposed methodology. It is shown that the volume fraction of the component used for dispersion strengthening by W_2N particles grows with increasing W content. According to field experiments, the cutting tool with the hardened coating has the increased wear resistance. The fatigue life of drill tools determined by the number of drilled holes increases by 2.2 times when drilling the type 30HGSA (ČSN 14331) steel and by more than 7 times when drilling the titanium alloy VT23.

The formation of secondary precipitates in spray-formed M3:2 steels with and without the addition of 2% Nb during 600 °C overtempering was described in [104], and precipitates were clearly identified by TEM. The steel hardening in the most quenched state occurred due to coherent precipitation of M_6C and MC carbides. During overtempering, the M3:2 steel hardness considerably lowered due to M_2C coarsening with the M_3C formation, while fine and dense M_2C precipitates were responsible for the excellent long-term strength of the Nb-containing steel. The Nb addition led to a significant dissolution of Mo and Cr in the matrix prior to quenching that enhanced the dispersion of secondary M_2C precipitate and retarded extended ageing. And the excellent tempering resistance could thus be obtained.

Using SEM, TEM, XRD, and EDS methods, Pan et al. [105] characterized the effect of austenitizing temperature on the carbide morphology and distribution and the martensite (Ms) content in high-speed cobalt steel. They found that eutectic MC carbide did not solute during austenitization, while lamellar M_2C carbide disintegrated into MC and M_6C carbides at 1100 °C. A large content of $M_{23}C_6$ carbides uniformly distributed in the matrix, dissolved in austenite at 1100 °C. At elevated temperature of austenitization, the alloying element dissolved in the matrix, and the solid solution strengthening of martensite increased, thereby enhancing the cobalt steel hardness. When the temperature of austenitization exceeded 1050 °C, the excess alloying elements in the matrix declined the Ms point and increased the volume fraction of residual Ms that resulted in a reduction of the cobalt steel hardness. The highest hardness of 66.4 HRC was detected at 1050 °C austenitization.

The effect of quenching temperature on the microstructure and properties of high-speed axle steel was investigated in [106] using SEM and TEM observations. These studies showed that with increasing quenching temperature, yield strength and tensile strength rapidly increased and then slowly decreased, plasticity changed insignificantly, and absorbed impact energy continuously reduced, while the primary austenitic grain and lath martensite grew. At 900 °C quenching, the austenitic grain rapidly coarsened, while carbides refined after successive tempering. The best complex properties the steel acquired after 850 °C quenching and 675 °C tempering, i.e., 796 MPa ultimate tensile strength, 677 MPa yield strength, 24.5% elongation, 82 and 72 J/m^2 absorbed energy at 25 and – 40 °C, respectively.

Wang et al. [107] used spray-formed vanadium-modified HSS for different heat treatment and investigated its microstructure, residual stress, and mechanical property evolution. They studied the effect of the quenching temperature on the steel hardness, tensile strength, and fracture toughness. The behavior of secondary carbides and their strengthening mechanisms were discussed together with TEM investigations. They proposed a mechanism of studying the carbide behavior during different heat treatment. The data obtained can be used to understand the microstructure and mechanical property evolution of spray-formed HSS and to optimize the alloy structure and post-heat treatment.

Jin et al. [108] gave a full description for the solidified microstructure and behavior of carbide precipitates in the HS steel S390 after electron beam melting (EBM). The obtained microstructure consisted of a discontinuous network of fine primary carbides dispersed in the self-tempering martensite matrix combined with a limited content of residual austenite. The carbide network comprised M_2C/M_6C and MC carbides. Both columnar and equiaxial grains were found in the microstructure, while the presence of interdendritic eutectic carbides helped to identify the dendritic nature of solidification. The upper microstrucutre layer showed that dendritic columnar grains locate near the micromolten pool that indicated to their epitaxial growth of the average direction parallel to the maximum temperature gradient. The formation of almost equiaxial grains at the center of the micromolten pool was stipulated by their dendritic growth parallel to the beam direction. Carbide decomposition was detected by SEM and confirmed by the transmission Kikuchi diffraction. MC carbides rich in V and W nucleated at the M_2C (order of priority: W, Fe, Mo, and Co)/matrix interface and then grew from outside to inside. The heat induced by neighboring scan lines probably initiated the solid-phase transformation $MC \rightarrow M_2C + \gamma$-Fe. The element migration was theoretically calculated and compared to the experimental data. The high (~ 65 HRC) hardness and good (~ 2500 MPa) tensile strength of the obtained S390 steel meant that EBM could be used to fabricate high alloyed tool steels. Thus, the heat treatment provided the steel hardness and tensile strength of 73.1 $\pm$ 0.2 HRC and 3012 $\pm$ 34 MPa, respectively.

Rahman et al. [109] used two HSS alloys for laser cladding onto a cylindrical substrate made of $42CrMo_4$ steel using a 4 kW Nd:YAG laser. After the optimization of processing parameters, multilayer claddings were obtained for single layers. SEM, EDS, EBSD, and high-resolution TEM were used to investigate carbides and the matrix. The alloy strengthening (Fe–Cr–Mo–W–V) was provided by the martensite matrix and residual austenite with eutectic carbide networks VC and Mo_2C. Fine carbides enriched with Cr (Cr_7C_3 and $Cr_{23}C_6$) were implanted in the matrix. During multilayer laser cladding, each successive layer had a negative effect on the hardness of already deposited layers that was explained by lath martensite tempering. In order to overcome the drop in hardness, a new alloy LC2 (Fe_{bal-x}–Cr–Mo–W–V–Co_x) was blended by the addition of 3–5% of Co in LC1. The Co addition provided the general hardness growth and its lower drop during successive laser cladding. The latter was conditioned by the Co presence in the Fe-based solid solution. High-resolution TEM characterized nanosized formations observed during reheating. These carbides were either rich in V and W or generated from a complex combination of V, Mo, W, and Cr having the lattice spacing of 0.15–0.26 nm.

The steel-type SA508-3 is widely used to produce reactor core thanks to its outstanding neutron radiation resistance and good fracture toughness. High-speed milling of this steel is studied in [110], and milling parameters of uncoated carbide tools are compared with TiAlN- and Al_2O_3-coated tools. It is shown that adhesive wear occurs more often in uncoated carbide tools, whereas the coating delamination is observed in TiAlN- and Al_2O_3-coated tools. Moreover, Al_2O_3-coated tools possess the best milling parameters than TiAlN-coated and uncoated tools, with respect to their service life and surface quality. The service life of Al_2O_3-coated carbide tool reaches 200 min, while the volume of material removed with a blunt tool is 182 $\times$ 10^3 mm^3. Experimental research into the tool life and wear promotes the higher milling quality and provides a theoretical basis for selection of the tool material and optimization of the SA508-3 steel milling process.

1.4 Experimental and Theoretical Studies of Electron Beam Processing Effect on the Structure and Properties of Alloys and Steels

As stated above, electron beam processing is one of the most promising methods of the surface modification of materials, including different steel grades with a view to enhance their mechanical and tribological properties [111]. During relatively short (~ 1–100 μs) time, the electron beam can generate such surface structures, depending on the chemical composition and process conditions, which provide the

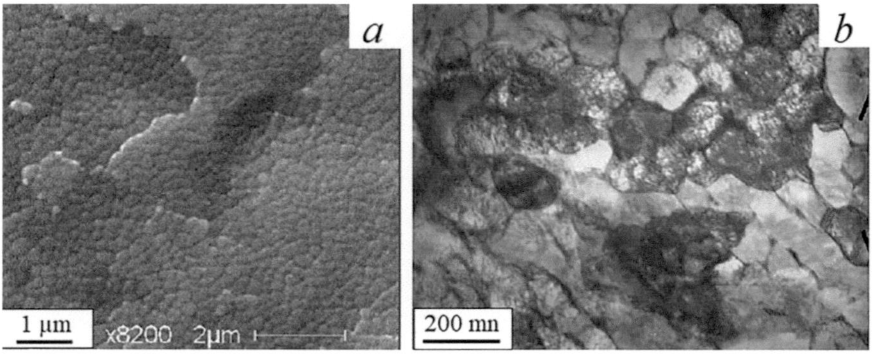

Fig. 1.8 SEM (**a**) and TEM (**b**) images of the type 20Kh13 steel structure with localized crystallization areas[1] observed in the surface layer after electron beam processing at 20 J/cm^2 [117]

high (~ 400–800 HV) microhardness, wear resistance, and fatigue life 2 or 3 times higher than that of unprocessed materials [112]. High-current electron beams have a complex effect on materials and include heating, thermoelastic wave propagation, melting, convective currents in a liquid layer, vaporization, and crystallization [113]. These processes determine the structure, phase composition, and properties of surface layers. According to [114], pseudospark pulsed electron beam of 100–1000 ns duration and 10^9 W/cm^2 power density leads to the surface modification of AISI 304 stainless steel after 5000 pulses, in which submicron-sized grains are then found. A fine structure of this layer represents nanotwins, subgrains, dislocation pile-ups, and cellular dislocation structures.

Stainless steel processing with ~ 100 μs electron beam with the power density of ~ 10^5–10^6 W/cm^2 also leads to the formation of the surface layer consisting of submicron- and nanosized grains [115–122]. Thus, the 20Kh13 steel processing with the electron beam with 20 J/cm^2 energy density provides the formation of the gradient multiphase structure. The surface layer ~ 5 μm thick is characterized by a complete dissolution of submicron-sized $M_{23}C_6$ particles in the initial steel, grains of two scale levels, multiphase cellular structure, $\gamma \rightarrow \varepsilon$ martensitic transformation, and large number of bend extinction contours [115, 116]. At 30 J/cm^2 energy density, submicron-sized $M_{23}C_6$ particles completely dissolve in the surface layer, α-phase grains reduce, and multiphase cellular structure appears [116]. These structural changes lead to the growth in the fatigue life of steel. Bessonov et al. [116] report that the high-current electron beam processing of 20Kh13 steel at 10, 20, 30 J/cm^2, results in 1.4, 1.6, and 1.9 times increase in the number of cycles to fracture. In this case, the dependence between the latter and the energy density is linear due to the grain and subgrain structure refinement, carbide phase dissolution, decrease in the Cr content in the surface layer, and cellular structure formation [117, 118]. Figure 1.8 presents SEM and TEM images of the 20Kh13 steel structure.

[1] Localized crystallization areas appear in the structure during rapid solidification.

Electron beam processing of the type 20Kh23N18 steel leads to an increased fatigue life, but the dependence between the number of cycles to fracture and the energy density is nonmonotonic [119]. The longest (a 2.1 times growth) fatigue life is observed at 20 J/cm^2 energy density. This is explained by the fact that these processing parameters do not lead to the critical structure formation, i.e., microcracks. According to SEM and TEM, the surface layer 3–5 μm thick starts to rapidly melt and solidify at 20 J/cm^2, and the grain size reduces from 41.4 to 21.3 μm. The grain volume consists of two or three microtwin systems and ~ 5.7 $\times$ 10^{10} cm^{-2} scalar dislocation density. At 30 J/cm^2, the depth of the molten and solidified layer is ~ 10 μm. The structure of ~ 18.4 μm grains consists of crystallization areas generated by the high-speed steel cooling. At 40 J/cm^2, the structure of ~ 22.4 μm grains consists of 300–400 nm crystallization areas, and their size does not depend on their location relative to the grain boundary. In addition, grains contain a banded substructure and microtwins [120–122].

The evolution of the 08Kh18N10T steel structure was studied in works [123–128] during electron beam processing, which modified the structure of the surface layer. It was characterized by a 1.5–2 times reduction in the average grain size, dendritic growth, complete dissolution of the initial carbide phase, and nanosized particles of the secondary phase [123]. When the energy density was increased from 15 to 25 J/cm^2, the modified structure manifested the lower scalar dislocation density, linear microtwin density and bend extinction contours due to a decrease in the cooling rate from 5.7 $\times$ 10^6 to 2 $\times$ 10^6 K/s. The heating of the surface layer up to pre-melting temperatures observed at 15 J/cm^2, resulted in the grain structure refinement. It was found that the average transverse and longitudinal grain size was 8.2 $\pm$ 3.0 and 18.0 $\pm$ 7.3 μm, respectively [124]. The growth in the energy density up to 25 J/cm^2 led to the dendritic growth with average grain size of 318 $\pm$ 66.3 nm. Secondary particles in this structure were observed mostly along the boundaries of crystallization areas and much less inside the areas. TiC and $Cr_{15}Fe_9$ carbide particles had a rounded shape and ranged in size from 20 to 45 nm. A dislocation substructure was chaotically distributed in crystallization areas (scalar dislocation density: 3.6 $\times$ 10^{10} cm^{-2}). Regardless of the layer thickness within 80 μm, the grain structure of the processed steel was characterized by microtwins and dislocations. Similar to high-chromium steels, electron beam processing of 08Kh18N10T steel was accompanied by the particle dissolution of the primary carbide phase ($M_{23}C_6$). Gorbunov et al. [125] reported that the average particle size and density reduced when approaching to the irradiated surface. These structure modifications explained the nonmonotonic dependence between the number of cycles to fracture and the energy density [126]. The longest fatigue life of 08Kh18N10T steel was observed at 25 J/cm^2, i.e., 3.5 times longer than that of the initial steel. As reported in [126, 127], in these processing conditions, the dissolution of $M_{23}C_6$ particles in the surface layer was the main mechanism of the fatigue life improvement for steel with low Cr content. Fatigue loading induced the $M_{23}C_6$ particle dissolution at a depth of ~ 10 μm, while at ~ 80 μm depth, they remained unchanged. Their average size was 151.7 $\pm$ 43.0 nm. Along with the $M_{23}C_6$ particle dissolution, fatigue loading was accompanied by strain ageing, which induced the generation of TiC particles in grains, whose average size increased with the distance from the front surface of specimens [128].

The electron beam surface modification of low-chromium steels is studied in [129–132]. Lu et al. [129] show that after scanning electron beam treatment, nano-sized carbides appear in the surface layers of the 5CrNiMo steel that enhances its hardness to 845.6 HV, wear resistance by 7 times, and reduces the wear loss from 1.4 to 0.2 mg. A correlation detected between the mechanical properties and surface nanocrystallization opens opportunities for the production of 5CrNiMo mold steel with exceptional properties. Wang and Hao [130] report that the $40CrNiMo_7$ steel processing with 3 J/cm^2 high-current pulsed electron beam at 2.5 μs pulse duration and 1 to 50 pulse number leads to a gradual carbide dissolution in ~ 4 μm thick molten zone with increasing number of pulses. Finally, the surface microstructure consists of the austenite layer 150 nm thick and the low content of martensite phases. After 15 pulses, the surface microhardness reaches 553 HV, while the wear rate reduces down to 1/3 of the initial state. Fu et al. [131] also gain a 2–3 times growth in the surface microhardness of the type 30CrMnSiA steel using a continuous electron beam. This growth is determined by the formation of a complex structure consisting of acicular martensite, upper and lower bainite, and insignificant content of lamellar martensite, while the substrate material consists of perlite and fine pearlite. The surface modification of 5CrNiMo steel is conducted by Qin et al. [132] using scanning electron beam assisted by ultrasonic vibration. This surface treatment provides the increase in the quantity of equiaxial grains. Their average size is 6.04 μm, i.e., ~ 23.5% lower than after electron beam processing (7.9 μm) without ultrasound.

The following works focus on the structure and property modification of tool steels achieved by electron beam processing. Continuous electron beam processing of the T10 steel surface provides the microhardness growth from 252 to 839 HV [133]. The steel microstructure becomes gradient. The molten zone, HAZ, and the substrate are investigated by electron microscopy. δ-ferrite and acicular martensite are observed in the molten zone, while martensite and cementite are found in the HAZ. The substrate structure consists of ferrite and perlite. Ormanova et al. [134] suggest the surface treatment of W320 tool steel by the electron beam, plasma nitriding, and electron beam again. The treatment combining electron beam and plasma nitriding provides a 760 kg/mm^2 hardness, while after the electron beam, plasma nitriding, and electron beam processing, the hardness is 718 kg/mm^2. There is good agreement between the calculated depth (100 μm) of the electron beam penetration and experimental data (117 μm). Selective electron beam melting with preliminary powder heating is used in [135] to obtain M2 high-speed steel without cracks. The steel structure consists of α-Fe phase grains smaller than 7 μm and fine-grained carbide generated by the high cooling rate of the molten pool that explains the high (~ 70 HRC) hardness and excellent tribological properties of the steel surface. Its wear is 13.7% lower than in quenched and tempered forged analogs.

The structure and properties of E76F rail steel after electron beam processing are investigated in [136–139]. According to [136, 137], at the energy density of 10 J/cm^2, the polycrystalline structure generates in the surface layer with the average grain size of ~ 5 μm. The subgrain structure represents cells observed in the grains. The average cell size is 330 nm. The real cell size ranges between 130 and 670 nm. The energy density increased up to 30 J/cm^2, provides the growth in the average grain size up to ~ 8.5 μm. In both cases, the cellular structure is observed inside the grains. At 20 J/cm^2, the average, smallest, and largest grain size is 352, 200, and 800 nm, respectively, while at 30 J/cm^2, these values are 427, 200, and 800 nm, respectively. Along with the cellular structure, the surface layer consists of nanosized (50–70 nm) martensite particles locating on the grain boundaries. There are also perlite and the so-called pseudoperlite grains inside which there occur different stages of thermal fracture of cementite lamellas and $\alpha \Rightarrow \gamma \Rightarrow \alpha$ transformation. Globular cementite particles ranging from 15 to 25 nm are observed at triple point intersections and along pseudoperlite grain boundaries. Inside these grains and subgrains, there are lath martensite crystals having a transverse size of 30–50 nm. Such structure and phase composition modifications have no choice but affect the mechanical properties of the steel, in particular fatigue life. The nonlinear dependence is obtained in [138, 139] for the number of cycles to fracture and the energy density of the electron beam. The highest point on this dependence matches 20 J/cm^2 (the number of cycles to fracture increased by 2.5 times). In these conditions, the formation of stress concentrators primarily occurs at the interface appearing between the rapid solidification zone and HAZ, as presented in Fig. 1.9 [139].

Electron beam processing provides the acicular interface profile, which leads to a dispersion of stress concentrators and promotes more homogeneous plastic yield in the substrate [140].

It can thus be concluded that electron beams with different energy density and pulse duration make the structure of the surface layer more complex and provide a well-defined gradient of the structure and phase transformations. In order to explain the gradient formation, it is necessary to simulate processes observed during the electron beam treatment. The gradient simulation of micro and nanostructured phase transformations is advisable to start from a search for the in-depth temperature distribution in the material.

Such process simulations are presented in works [141–145] with respect to first-order phase transitions (melting, evaporation). Mirzoev et al. [141] presented the laser irradiation effect on the target with its evaporation and vapor dispersion. This problem is a complex gas-dynamic flotation in a flame of the evaporated material and thermal processes in the target. The melt front is, however, not studied properly here. In works [142, 143], the model includes the temperature field and evaporation rate during magnetron sputtering of the target. These parameters are calculated from the boundary value problem for heat equation with regard to the evaporation rate. The Hertz–Knudsen equation is used to calculate the evaporation rate, but processes in the target are not analyzed. Khaimzon et al. [144] propose the model considering the kinetic equation for particles, continuum mechanics equations, and wide ranging equation of state for the temperature distribution produced by pulsed energy fluxes with the target evaporation. Note that the dependence between the penetration depth and beam parameters is not investigated. It is therefore advisable to use the simplified model proposed in [144], which properly describes temperature fields in the target and allows to identify the penetration depth.

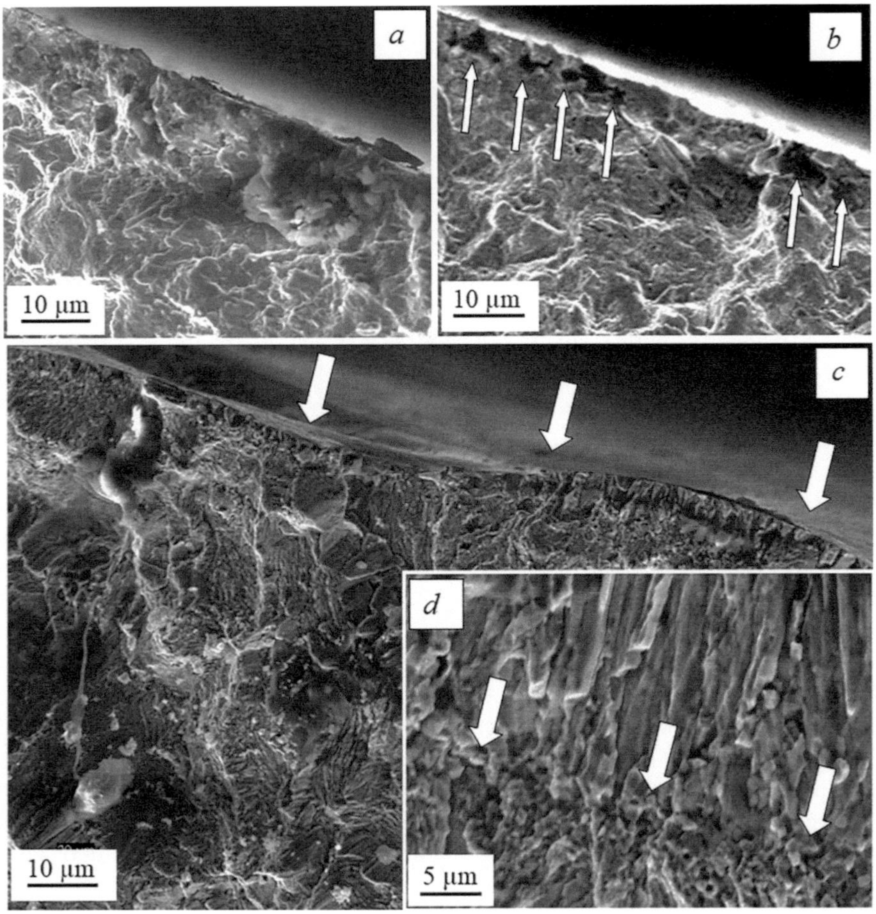

Fig. 1.9 SEM images of E76F rail steel surface [138, 139]: **a**—initial state, **b**—after electron beam processing at 10 J/cm^2, **c**, **d**—20 J/cm^2. Arrows indicate pores in the surface layer (**b**), irradiated surface (**c**), transition layer between solidification zone and HAZ (**d**)

Slyadnikov et al. [145] studied the melting kinetics in using low-energy high-current electron beams. They solved the system of heat-transfer equations and the Ginzburg–Landau equation. The ordering parameter in that model was a measure of deviation of the atomic configuration in less symmetrical phase from the atomic configuration in more symmetrical phase. It was found that the first-order phase transition induced by the volume heat source, occurred in nonequilibrium conditions of a significant localized volume overheating, and was accompanied by the formation of localized unstable state of the single crystal. The phase transition involved both surface and volume mechanisms. At low heating rates, the melting process was described by the system of two fronts—ordering parameter and temperature—propagating in space at irregular velocity (at first, acceleration proportional to the time root and then restricted velocity). The single crystal in a metastable state was stable relative to small fluctuations. If a high critical fluctuation of a new phase occurred on its free lateral surface, melting initiated and propagated to the center of the single crystal in the form of a transition from the metastable critical state to stable liquid state, i.e., the surface phase transition occurred. In the presence of the grain boundary in the crystal, melting initiated on both the free surface and grain boundary. The melt front comprised four atomic layers, and the rate of the melt front propagation was not constant for the surface phase transition. It was proportional to the heat intensity and restricted by physical parameters of the medium. The system of heat-transfer equations was solved in the assumption that coefficients of the heat capacity, heat transfer, and substance density were constant.

Real coefficients depend on time and temperature, and hence, the proposed model gives an idea only about the melting process of metal during electron beam processing. Thermal mathematical models give an idea only about the temperature distribution in a specimen, and this is important information for creating models of thermoelastic wave propagation, thermocapillary melt flow, and refinement of the secondary-phase particles in alloys.

As mentioned above, electron beam processing leads to the formation of a cellular structure on the steel surface due to the melt flow instability. Bugaev et al. [146] investigated the thermocapillary instability of this structure and obtained the dispersion relation of the solid–liquid boundary perturbations, which determined the wavelength of the most rapidly grown perturbation mode λ_m. That approach to electron beam processing showed that the λ_m value ranges between 50 and 200 μm, whereas its nanometer range was within 10^{10}–10^{12} K/m [147]. These results meant that thermocapillary instability did not fully explained the formation of the cellular structure. Therefore, it is expedient to consider other instabilities such as evaporation-capillary and thermoelectric instabilities. The evaporation-capillary instability was investigated in [148] in two evaporation modes: subsonic [148] and evaporation into vacuum or low back pressure medium [149]. Laser-induced instabilities considered in [150] were conditioned by the incident radiation diffraction on initial roughness. Diffracted and pumping wave interference led to periodic temperature changes, resulting in the formation of thermocapillary forces and recoil force, which amplified perturbations. Such interference instabilities provided the formation of ordered surface texture with a period on the order of a wavelength and were typical for only pulsed laser irradiation. Works of [151, 152] were devoted to thermocapillary instabilities with regard to evaporation. In these works, longwave approximation was used in the case when the flow analysis was reduced to one equation of the layer thickness.

Thermoelectric instability at concentrated energy fluxes is studied in [153–155]. They solve Pearson's square problem for top and bottom heating. The criterion of thermoelectric instability is determined as $E = \frac{\varepsilon\delta^2 G^2 h^2}{\rho\nu\kappa}$, where ε is the dielectric permittivity, $G = |\nabla T|$ is the temperature gradient modulus, h is the layer thickness, κ is the thermal conductivity, and δ is the thermoelectric constant. In the physical sense, this value shows the relation between the electric and dissipative forces. In the case of pure thermoelectric perturbation, the instability occurs at $E > 40$ [153]. A quality thermoelectric perturbation in a liquid can be understood as follows. The temperature fluctuation causes the perturbation of the thermoelectric field, which induces the formation of the space charge. And the space charge is affected by the electric force in the outer thermoelectric field. The electric force sets in motion the liquid if it exceeds the dissipation force by the required number (E) of times [154]. According to [155], thermoelectric instability becomes significant at a large temperature gradient (~ 10^6–10^{12} K/m) and thickness range (10–100 μm) of molten layers. The formation of micro and nanosized structures during electron beam processing can therefore be stipulated by the combined thermal, concentration-capillary, evaporation-capillary, and thermoelectric instability. The linear analysis of this combination proposed in [156–159] shows that in processing the binary Ti-Y alloy with the electron beam at 35 J/cm^2 energy density and 150 μs pulse duration, the dependence between the growth rate of perturbations at the liquid/plasma interface and the wavelength has one peak at a 48 μm, with respect to concentration-capillary and thermoelectric instabilities. With consideration to the evaporation-capillary instability, this wavelength reduces to 36 μm. At the thermoelectrical coefficient ~ 10^{-1} V/K, the peak growth rate occurs in the submicro- and nanosized wavelength range [156, 157]. The same effect is observed in [158, 159] for high entropy alloys at ~ 4 V/K only. According to the linear analysis of stability, the wavelength matching the largest perturbation mode reduces with increasing energy density, while the experimental values of crystallization areas grow. This can be explained by an increase in evaporation losses at a higher energy density that leads to a reduction in the nonperturbated temperature gradient [159] and, consequently, λ_m value (Fig. 1.10).

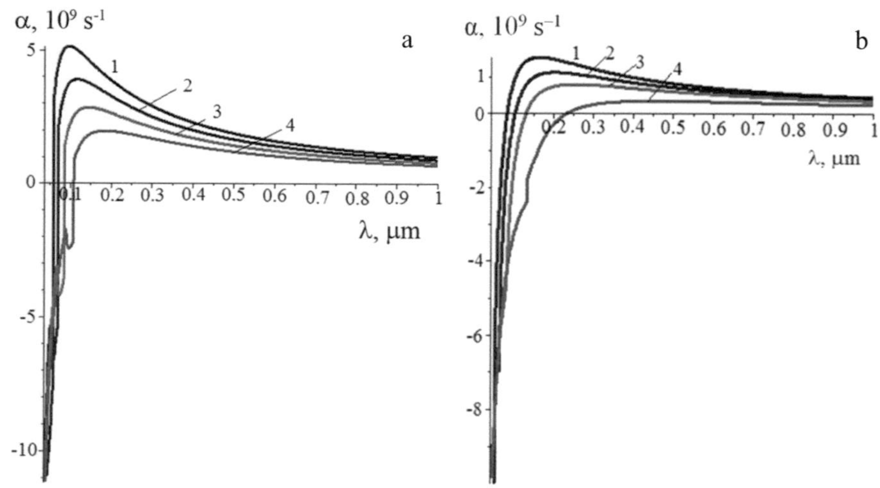

Fig. 1.10 Dependences between the growth rate of perturbations on AlCoCrFeNi (**a**) and CoCrFeMnNi (**b**) alloy surface and wavelength at 30 J/cm^2 and different evaporation losses: 1—0.1, 2—0.2, 3—0.3, 4—0.4 [159]

According to the literature review, electron beam processing of different steel grades is a complex and multifactorial process. Despite the different chemical composition of these steels, the electron beam generates a modified surface layer consisting of the melting zone and HAZ. High hardness values are attributed to the formation of micro and nanostructure, decomposition of carbide phases, and the formation of carbon-containing oversaturated solid solutions in α- and γ-iron, depending on the steel chemical composition. This literature review also shows that there are currently no models of the electron beam irradiation, clearly describing the formation of gradient structures and phase compositions of metallic materials having different chemical compositions.

1.5 Conclusions

The literature review in this area highlighted the problem of improving the reliability and durability of mechanisms and machinery, which is being solved today by goal-oriented and controlled changes in their properties and working surfaces.

Mining, construction, and metal industries utilize plasma-jet hard-facing to protect metals and alloys from different external influences such as wear, corrosion, and static and dynamic loads and to provide them with a set of high mechanical properties. Today, the foremost problem is to fabricate coatings with high-performance properties, ensuring higher reliability and longer life time of products in extreme conditions (high mechanical loads, wear, corrosion environments, and thermal cycling).

The coating materials differing from the base material of parts and providing the appropriate surface properties is the most important determinant of steel hardening. High-speed steels are often used for these purposes.

Plasma surfacing in shielding gas (nitrogen) is the most cost effective and efficient technique of the surface protection. It provides the best ratio between the surface properties and the bulk material. Plasma surfacing is used not only to repair worn structural elements, but also to give specific characteristics to the surface of new products prior to startup.

Justification of the selection of the hard-facing material used in certain operational conditions requires intensive research into the structure, phase composition, mechanical and tribological properties, and their evolution during the successive heat treatment.

The additional surface strengthening can be obtained via electron beam processing, one of the most promising methods of metallic material modification, which leads to a significant improvement of mechanical properties of the entire material due to its surface structural optimization.

References

1. H.F. Fischmeister, A.D. Ozerskii, L. Olsson, Solidification structure of gas-atomized high-speed steel powders. Powder Metall. **25**, 1–9 (1982)
2. H.F. Fischmeister, R. Riedl, Solidification of high-speed tool steels. Metall. Mater. Trans. **20**, 2133–2148 (1989)
3. F. Meurling, A. Melander, M. Tidesten, L. Westin, Influence of carbide and inclusion contents on the fatigue properties of high speed steels and tool steels. Int. J. Fatig. **23**, 215–224 (2001)
4. J.W. Park, H.C. Lee, S. Lee, Composition, microstructure, hardness, and wear properties of high-speed steel rolls. Metall. Mater. Trans. **30**, 399–409 (1999)
5. E. Badisch, C. Mitterer, Abrasive wear of high speed steels: influence of abrasive particles and primary carbides on wear resistance. Tribol. Int. **36**, 765–770 (2003)
6. M. Domankova, Microstructure and properties of M2 high-speed steel cast by the gravity and vacuum investment casting. Vacuum **162**, 183–198 (2019)
7. H. Fu, Q. Xiao, J.D. Xing, A study on the crack control of a high speed steel roll fabricated by a centrifugal casting technique. Mater. Sci. Eng. A **474**, 82–87 (2008)
8. X. Zhou, F. Fang, J.Q. Jiang, A study on the microstructure of AISI M2 high speed steel manufactured by continuous casting. Adv. Mat. Res. **146–147**, 1211–1215 (2011)
9. H. Halfa, Characterization of electroslag remelted super hard high speed tool steel containing niobium. Steel Res. Int. **84**, 495–510 (2013)
10. Z. Ximin, D. Xin, J. Zhouhua, Q. Tianyu, L. Chunxu, New manufacture of high speed steel 90 mm billets by electroslag remelting withdrawing process. Iron Steel **1**, 39–45 (2016)
11. Y.W. Luo, H.J. Guo, X.L. Sun, J. Guo, F. Wang, Influence of tempering time on the microstructure and mechanical properties of AISI M42 high-speed steel. Metall. Mater. Trans. **49A**, 5976–5986 (2018)
12. Y.H. Moon, J.W. Kim, D.W. Lee, Machining characteristics of electroslag cast steel for hot-working tools. J. Mater. Process. Technol. **153–154**, 654–659 (2004)
13. H. Peng, L. Hu, L. Li, L. Zhang, X. Zhang, Evolution of the microstructure and mechanical properties of powder metallurgical high-speed steel S390 after heat treatment. J. Alloys Compd. **740**, 766–773 (2018)
14. W. Shen, L. Yu, H. Liu, Diffusion welding of powder metallurgy high speed steel by spark plasma sintering. J. Mater. Process. Technol. **275**, 1–9 (2020)
15. H. Peng, L. Hu, X. Zhang, X. Wei, L. Li, J. Zhou, Microstructural evolution, behavior of precipitates, and mechanical properties of powder metallurgical high-speed steel S390 during tempering. Metall. Mater. Trans. **50**, 874–883 (2018)
16. Y. Liu, Y. Ning, Z. Yao, M.W. Fu, Hot deformation behavior of the 1.15C–4.00Cr–3.00V–6.00W–5.00Mo powder metallurgy high speed steel. Mater. Des. **54**, 854–863 (2014)
17. R.A. Mesquita, C.A. Barbosa, High-speed steels produced by conventional casting, spray forming and powder metallurgy. Mater. Sci. Forum **498–499**, 244–250 (2005)
18. A.S. Chaus, J. Chovanec, M. Legerska, Development of high speed steels for cast metal-cutting tools. Solid State Phenom. **113**, 559–564 (2006)
19. P. Ding, S. Zhou, F. Pan, J. Liu, Ecotechnology for high-speed tool steels. Mater. Des. **22**, 137–142 (2001)
20. F. Yin, L. Wang, Z. Xiao, J. Feng, L. Zhao, Effect of titanium and rare earth microalloying on microsegregation, eutectic carbides of M2 high speed steel during ESR process. J. Rare Earths **38**, 1030–1038 (2020)
21. Q. Li, Z. Xia, Y. Guo, Z. Shen, T. Zheng, B. Ding et al., Carbides modification and mechanical properties enhancement of $Cr_{12}MoV$ die steel by magnetically controlled electroslag remelting. Metall. Mater. Trans. B **52**, 1495–1507 (2021)
22. S. Wan, H. Li, K. Tieu, Q. Xue, H. Zhu, Mechanical and tribological assessments of high-vanadium high-speed steel by the conventional powder metallurgy process. Int. J. Adv. Manuf. Technol. **103**, 943–955 (2019)
23. J. Edgar, S. Tint, Additive manufacturing technologies: 3D printing, rapid prototyping, and direct digital manufacturing, 2nd edn. Johnson Matthey Technol. Rev. **59**, 193–198 (2015)

24. L.E. Murr, S.M. Gaytan, D.A. Ramirez, E. Martinez, J. Hernandez, K.N. Amato et al., Metal fabrication by additive manufacturing using laser and electron beam melting technologies. J. Mater. Sci. Technol. **28**, 1–14 (2012)
25. R.A. Mesquita, C.A. Barbosa, Spray forming high-speed steel properties and processing. Mater. Sci. Eng. A **383**, 87–95 (2004)
26. A. Kg, A. Ar, B. Ffa, A. Wt, Microstructure, mechanical, and tribological properties of M3:2 high-speed steel processed by selective laser melting, hot-isostatic pressing, and casting. Addit. Manuf. **28**, 585–599 (2019)
27. N.U. Rahman, L. Capuano, S. Cabeza, M. Feinaeugle, Directed energy deposition and characterization of high-carbon high speed steels. Addit. Manuf. **30**, 100838–100848 (2019)
28. J. Saewe, C. Gayer, L. Jauer, J. Kunz, Feasibility investigation for laser powder bed fusion of high-speed steel. HS 6–5–3–8. EuroPM (2018)
29. J. Saewe, M.B. Wilms, L. Jauer, J.H. Schleifenbaum, Influence of preheating temperature on hardness and microstructure of high-speed steel HS6-5-3-8 manufactured by laser powder bed fusion, in *Proceedings of International Conference on Metal Additive Manufacturing* (2019)
30. J. Saewe, C. Gayer, A. Vogelpoth, J.H. Schleifenbaum, Feasibility investigation for laser powder bed fusion of high-speed steel AISI M50 with base preheating system. Berg Huettenmaenn Monatsh **164**, 101–107 (2019)
31. J. Saewe, N. Carstensen, P. Kurnsteiner, E.A. Jgle, J.H. Schleifenbaum, Influence of increased carbon content on the processability of high-speed steel HS6-5-3-8 by laser powder bed fusion. Addit. Manuf. **46**, 102125 (2021)
32. J. Jin, R. Gao, H. Peng, H. Guo, B. Chen, Rapid solidification microstructure and carbide precipitation behavior in electron beam melted high-speed steel. Metall. Mater. Trans. **51**, 2411–2429 (2020)
33. N.U. Rahman, L. Capuano, M. Rooij, D. Matthews, Laser metal deposition of vanadium-rich high speed steel: microstructural and high temperature wear characterization. Surf. Coat. Technol. **364**, 115–126 (2019)
34. J. Arias, M. Cabeza, G. Castro, I. Feijoo, P. Merino, G. Pena, Modification of AISI M2 high-speed tool steels after laser surface melting under different operating conditions. Weld. Int. **27**, 1–9 (2013)
35. J. Arias, M. Cabeza, G. Castro, I. Feijoo, P. Merino, G. Pena, Microstructural characterization of laser surface melted AISI M2 tool steel. J. Microsc. **239**, 184–193 (2010)
36. D.S. Shim, G.Y. Baek, E.M. Lee, Effect of substrate preheating by induction heater on direct energy deposition of AISI M4 powder. Mater. Sci. Eng. A **682**, 550–562 (2017)
37. N.U. Rahmana, L. Capuanoa, A. Van der Meera, M.B. De Rooijb, Development and characterization of multilayer laser cladded high speed steels. Addit. Manuf. **24**, 76–85 (2018)
38. E.S. Lee, W.J. Park, J.Y. Jung, S. Ahn, Solidification microstructure and M2C carbide decomposition in a spray formed high-speed steel. Metall. Mater Trans. **29**, 1395–1404 (1998)
39. E.S. Lee, W.J. Park, K.H. Baik, S. Ahn, Different carbide types and their effect on bend properties of a spray-formed high speed steel. Scripta Mater. **39**, 1133–1138 (1998)
40. L. Lu, L.G. Hou, J.X. Zhang, H.B. Wang, H. Cui, J.F. Huang et al., Improved the microstructures and properties of M3:2 high speed steel by spray forming and niobium alloying. Mater. Char. **117**, 1–8 (2016)
41. H. Fayazfar, M. Salarian, A. Rogalsky, D. Sarker, P. Russo, V. Paserin et al., A critical review of powder-based additive manufacturing of ferrous alloys: process parameters, microstructure and mechanical properties. Mater. Des. **144**, 98–128 (2018)
42. P. Bajaj, A. Hariharan, A. Kini, P. Kurnsteiner, E. Jagle, Steels in additive manufacturing: a review of their microstructure and properties. Mater. Sci. Eng. A **772**, 138633 (2020). https://doi.org/10.1016/j.msea.2019.138633
43. N. Haghdadi, M. Laleh, M. Moyle, S. Primig, Additive manufacturing of steels: a review of achievements and challenges. J. Mater. Sci. **56**, 64–107 (2021)
44. Y. Yin, Q. Tan, M. Bermingham, N. Mo, J. Zhang, Laser additive manufacturing of steels. Int. Mater. Rev. **67**(5), 487–573 (2022)

45. D.D. Gu, W. Meiners, K. Wissenbach, R. Poprawe, Laser additive manufacturing of metallic components: materials, processes and mechanisms. Int. Mater. Rev. **57**, 133–164 (2012)
46. S.M. Thompson, L. Bian, N. Shamsaei, A. Yadollahi, An overview of direct laser deposition for additive manufacturing; Part I: transport phenomena, modeling and diagnostics. Addit. Manuf. **8**, 36–62 (2015)
47. A. Martinez-Garcia, M. Monzon, R. Paz, Chapter 12—Standards for additive manufacturing technologies: structure and impact, in *Additive Manufacturing*, ed. by J. Pou, A. Riveiro, J.P. Davim (Elsevier, Amsterdam, 2021), pp. 395–408
48. E. Houdremont, in *Handbuch der Sonderstahlkunde*, vol. 2, ed. by A.S. Zaimovskii, M.L. Bernshtein (Metallugiya, Moscow, 1966) (Russian translation)
49. A.P. Gulyaev, *General Metallurgy* (Metallugiya, Moscow, 1986). (in Russian)
50. L.S. Kremnev, Y.A. Geller, Effect of tungsten on properties of high-speed steels, Izv, AN SSSR, Metallurgiya i gornoe delo, no. 1, pp. 136–142 (1964). (in Russian)
51. Y.A. Geller, Modern high-speed steels. Metal Sci. Heat Treat. **19**(10), 867–873 (1977)
52. I. Artinger, *Tool steels and their heat treatment: Handbook [Russian translation]* (Metallugiya, Moscow, 1982)
53. Yu.A. Geller, *Tool Steels* (Metallugiya, Moscow, 1983). (in Russian)
54. H.J Goldschmidt, *Interstitial Alloys* (Mir, Moscow, 1971) (Russian translation)
55. L.S. Kremnev, V.A. Brostrem, Thermal resistance of tool steels and alloys, MiTOM **3**, 46–52 (1973)
56. G.R. Speich, W.C Leslie, Tempering of steel. Metall. Trans. **3**, 1043–1053 (1972)
57. L.S. Kremnev, Yu.E. Sedov, Optimization of the compositions of low-alloy high-speed steels. Met. Sci. Heat Treat. **30**(6), 429–437 (1988)
58. L.S. Kremnev, Theory of alloying of high speed steels. Met. Sci. Heat Treat. **33**(6), 428–434 (1991)
59. A.N. Popandopulo, V.I. Kalinina, Carbon content in tungstenless powdered high speed steels. Met. Sci. Heat Treat. **32**, 229–231 (1990). https://doi.org/10.1007/BF00730931
60. A.S. Chaus, I.V. Latyshev, Effect of vanadium, titanium, and niobium on the structure and properties of cast tungsten-molybdenum high-speed steel. Phys. Met. Metallogr. **88**(5), 462–468 (1999)
61. W.-C. Jiao, H. Li, H. Feng, H.-J. Wang, H. Zhu, S.-C. Zhang, Z.-H. Jiang, W. Wu, Improved high-temperature softening resistance and wear resistance of AISI M42 high-speed steel by N-alloying. Steel Res. Int. **94**(8), 1–11 (2023)
62. Y. Li, C. Chen, J. Lv, H. Yang, S. Ma, M. Sun, Z. Jiang, The effect of magnesium treatment on the inclusions, microstructure, and mechanical properties of M7 high-speed steel. J. Mater. Eng. Perform. 1–18 (2023). https://doi.org/10.1007/s11665-023-08282-w
63. V.P. Kireev, HSS hardening by dynamic microalloying and its effect on wear resistance of cutting tool. Izv. Samarskogo nauchnogo tsentra Rossiiskoi akademii nauk **17**, 6(2), 414–418 (2015)
64. H.Y. Bor, C.G. Chao, C.Y. Ma, The influence of Mg on creep properties and fracture behaviors of Mar-M247 superalloy under 1255 K/200 MPa. Metall. Mater. Trans. A **31**(5), 1365–1373 (2000)
65. L.G. Petrova, A.S. Sergeeva, V.M. Vdovin, Surface modification of high-speed cutting tools by joint saturation with tungsten and nitrogen. Naukoemkie tekhnologii v mashinostroenii **7**(145), 24–32 (2023)
66. A.S. Akhmetov, Zh.V. Eremeeva, Microstructure and hardness modification after heat treatment of sintered high-speed steel R6M5K5 containing diffusion-alloyed additive. Perspektivnye materialy **3**, 43–48 (2023)
67. B.K. Rakhadilov, W. Wieleba, M.K. Kylyshkanov, A.B. Kenesbekov, M. Maulet, Structure and Phase composition of high-speed steels. Bulletin of the Karaganda University. Phys. Ser. **2**(98), 83–92 (2020)
68. H. Wang, L. Hou, Y. Li, P. Ou, L. Shen, X. Wen, H. Cui, J. Zhang, Effect of niobium on the secondary precipitates and tempering resistance of spray-formed M3:2 high-speed steel. J. Mater. Eng. Perform. **28**, 926–937 (2019)

69. Y. Wang, S. Chu, B. Mao, H. Xing, J. Zhang, B. Sun, Microstructure, residual stress, and mechanical property evolution of a spray-formed vanadium-modified high-speed steel processed by post-heat treatment. J. Mater. Res. Technol. **18**, 1521–1533 (2022). https://doi.org/10.1016/j.jmrt.2022.03.053
70. Q. Ouyang, P. Luo, F. Zhang, Q. He, Y. Wang, S. Li, Analyzing the effect of CeB_6 on microstructure and mechanical properties of high-speed steel consolidated by powder metallurgy. J. Mater. Eng. Perform. **27**, 5973–5983 (2018)
71. A.S. Chaus, P. Pokorný, Ľ Čaplovič, M.V. Sitkevich, J. Peterka, Complex fine-scale diffusion coating formed at low temperature on high-speed steel substrate. Appl. Surf. Sci. **437**, 257–270 (2018)
72. M.G. Krukovich, B.A. Prusakov, I.G. Sizov, Plasticity of boronized layers. Springer Ser. Mater. Sci. **237**, 364 (2016)
73. B. Edenhofer, D. Joritz, M. Rink, K. Voges, Carburizing of steels, in *Thermochemical Surface Engineering of Steels*, ed. by E.J. Mittemeijer, M.A.J. Somers (Woodhead Publishing, 2015), pp. 485–553
74. E.D. Doyle, A.M. Pagon, P. Hubbard, S.J. Dowey, A. Pilkington, D.G. McCulloch, Nitriding of high speed steel. Int. Heat Treat. Surf. Eng. **5**, 69–72 (2011)
75. A.S. Chaus, M. Murgas, I.V. Latyshev, R. Toth, Heat treatment of cast carburising high-speed steel alloyed with Ti, Nb and V. Met. Sci. Heat Treat. **43**, 220–223 (2001)
76. A.P. Gulyaev, K.A. Malinina, S.M. Saverina, *Tool Steels: Reference book* (Mashinostroenie, Moscow, 1975). (in Russian)
77. S.-H. Chang, T.-C. Tang, K.-T. Huang, C.-M. Liu, Investigation of the characteristics of DLC films on oxynitriding-treated ASP23 high speed steel by DC pulsed PECVD process. Surf. Coat. Technol. **261**, 331–336 (2015)
78. W. Wu, W. Chen, S. Yang, Y. Lin, S. Zhang, T.Y. Cho, G.H. Lee, Kwon SC. Design of AlCrSiN multilayers and nanocomposite coating for HSS cutting tools. Appl. Surf. Sci. **351**, 803–810 (2015)
79. I.S. Cho, A. Amanov, J.D. Kim, The effects of AlCrN coating, surface modification and their combination on the tribological properties of high speed steel under dry conditions. Tribol. Int. **81**, 61–72 (2015)
80. D. Kottfer, M. Ferdinandy, L. Kaczmarek, I. Maňková, J. Beňo, Investigation of Ti and Cr based PVD coatings deposited onto HSS Co 5 twist drills. Appl. Surf. Sci. **282**, 770–776 (2013)
81. N. Hashemi, A. Mertens, H.-M. Montrieux, J.T. Tchuindjang, O. Dedry, R. Carrus, J. Lecomte-Beckers, Oxidative wear behaviour of laser clad high speed steel thick deposits: Influence of sliding speed, carbide type and morphology. Surf. Coat. Technol. **315**, 519–529 (2017)
82. G.F. Sun, K. Wang, R. Zhou, A.X. Feng, W. Zhang, Effect of different heat-treatment temperatures on the laser cladded M3:2 high-speed steel. Mater. Des. **65**, 606–616 (2015)
83. C. Soriano, J. Leunda, J. Lambarri, N.V. García, C. Sanz, Effect of laser surface hardening on the microstructure, hardness and residual stresses of austempered ductile iron grades. Appl. Surf. Sci. **257**, 7101–7106 (2011)
84. A.S. Chaus, Ĺ Čaplovič, A.V. Maksimenko, N.N. Fedosenko, V.N. Myshkovets, Formation of structure of an annealed high-speed steel upon laser surface melting. Phys. Met. Metallogr. **120**(4), 371–377 (2019)
85. V.A. Korotkov, I.A. Rastegaev, D.L. Merson, M.A. Afanas'ev, Influence of plasma thin-film coating of Si–O–C–N system on high-speed steel surface hardening. Poverkhnost'. Rentgenovskie, sinkhrotronnye i neitronnye issledovaniya **3**, 62–70 (2020)
86. D.V. Mamontov, I.D. Sklizkov, A.V. Asylbaev, R.K. Vafin, Influence of severe plastic deformation on hardened layer properties of R6M5 tool high-speed steel. Mater. Technol. Design **4**, 1(7), 30–38 (2022)
87. R.Z. Valiev, I.V. Aleksandrov, *Nanostructured Materials Obtained by Severe Plastic Deformation* (Logos, Moscow, 2000). (in Russian)
88. M.S. Migranov, A.M. Migranov, S.M. Minigaleev, S.R. Shehtman, Tribological properties of multilayer coatings for cutting tool. J. Frict. Wear **39**, 245–250 (2018)

89. M. Migranov, R. Migranova, Tool coatings with the effect of adaptation to cutting conditions. Key Eng. Mater. **496**, 75–79 (2012)
90. D.I. Tokarev, A.A. Drozdov, E.A. Morozov, T.R. Ablyaz, Y.S. Ivanova, I.S. Polezhaev, Influence of laser treatment on hardened high-speed steel R6M5. Vestnik PNIPU. Mashinostroenie, materialovedenie **21**(4), 34–41 (2019)
91. A.D. Kuliev, Technological processes of laser processing. Ph.D Thesis. Moscow (1983)
92. Mitrofanov A. A. Increasing the performance of high-speed cutting tools by laser technology. PhD. Thesis, Moscow (1998)
93. I.K. Kupalova, Structural hardening of high-speed steel. Phys. Met. Metallogr. **123**(9), 933–938 (2022)
94. A.S. Chaus, Structure and phase transformations in carbides after thermal treatment of high-speed steel. Fiz. Met. Metalloved. **117**(7), 706–714 (2016)
95. Y. Luan, N. Song, Y. Bai, X. Kang, D. Li, Effect of solidification rate on the morphology and distribution of eutectic carbides in centrifugal casting high speed steel rolls. J. Mater. Process. Technol. **210**, 536–541 (2010)
96. M. Boccalini, H. Goldenstein, Solidification of high speed steels. Int. Mater. Rev. **46**(2), 92–115 (2001)
97. A.S. Chaus, Microstructural and properties evaluation of M2 high speed steel after inoculating addition of powder W and WC. Mater. Sci. Technol. **30**(9), 1105–1115 (2014)
98. A.N. Zhigalov, I.A. Goravskii, Microstructure of high-speed steel R6M5 strengthened by aerodynamic sound hardening. Vestnik Baranovichskogo gosudarstvennogo univ. Ser.: Tekhnich. nauki **1**(11), 14–23 (2022)
99. D.I. Tokarev, A.A. Drozdov, E.A. Morozov, T.R. Ablyaz, Y.S. Ivanova, I.S. Polezhaev, Laser treatment of hardened high-speed steel R6M5. Vestnik Permskogo natsional'nogo issledovatel'skogo politekhnich. univ. Mashinostroenie, materialovedenie **21**(4), 34–41 (2019)
100. V.Y. Popov, D.A. Rychkov, P.V. Arkhipov, A.M. Kuznetsov, E.D. Losev, N.V. Selin, Surface quality of HSS tools after electric powered diamond grinding. Obrabotka metallov: tekhnologiya, oborudovanie, instrumenty **21**(1), 6–15 (2019)
101. A.Y. Lavrent'ev, A.M. Dozhdelev, Improvement of heat affected zone structure of composite metal tools. Nauchno-tekhnicheskie vedomosti SPbPU. Estestvennye inzhenernye nauki **23**(3), 118–126 (2017)
102. N.N. Cherenda, V.V. Uglov, A.M. Kashevskii, V.M. Astashinskii, A.M. Kuz'-mitskii, Modification of phase composition and microhardness of HSS surface layer by combined plasma and thermal processing. Zhurnal Belorusskogo gosudarstvennogo universiteta. Fiz. **2**, 61–70 (2018)
103. L. Petrova, A. Sergeeva, V. Vdovin, Modification of a high-speed cutting tool surface by combined tungsten steel and nitrogen saturation. Sci. Intensive Technol. Mech. Eng. **2023**(7), 24–32 (2023). https://doi.org/10.30987/2223-4608-2023-7-24-32
104. H. Wang, L. Hou, Y. Li, P. Ou et al., Effect of niobium on the secondary precipitates and tempering resistance of spray-formed M3:2 high-speed steel. J. Mater. Eng. Perform. **28**, 926–937 (2019)
105. Y. Pan, T.N. Jin, N. Yuan, T. Ma, H. Fu, Effect of austenitizing temperature on microstructure and properties of a high-speed cobalt steel. Mater. Test. **64**(8), 1122–1135 (2022). https://doi.org/10.1515/mt-2021-2187
106. C.-C. Ding, M.-D. Zhao, Z.-D. Li, Y.-G, Cao Effect of quenching temperature on microstructure and properties of high-speed axle steel. Cailiao Rechuli Xuebao/Trans. Mater. Heat. Treat. **39**(12), 49–56 (2018). https://doi.org/10.13289/j.issn.1009-6264.2018-0311
107. Y. Wang, S. Chu, B. Mao, H. Xing et al., Microstructure, residual stress, and mechanical property evolution of a spray-formed vanadium-modified high-speed steel processed by post-heat treatment. J. Mater. Res. Technol. **18**, 1521–1533 (2022)
108. J. Jin, R. Gao, H. Peng, H. Guo et al., Rapid solidification microstructure and carbide precipitation behavior in electron beam melted high-speed steel. Metall. Mater. Trans. A Phys. Metall. Mater. Sci. **51**(5), 2411–2429 (2020). https://doi.org/10.1007/s11661-020-05661-z

109. N. Rahman, L. Capuano, A. van der Meer, M. de Rooij et al., Development and characterization of multilayer laser cladded high speed steels. Add. Manuf. **24**, 76–85 (2018). https://doi.org/10.1016/j.addma.2018.09.009
110. W. Qinqiang, Y. Zhao, X. Zhu, J. Zhang et al., Effect of milling parameters on machinability of SA508-3 steel in high-speed milling with uncoated and coated carbide tools. Proc. Inst. Mech. Eng. B: J. Eng. Manuf. **238**(8) (2023). https://doi.org/10.1177/09544054231189604
111. V.A. Shulov, A.N. Gromov, D.A. Teryaev, V.I. Engel'ko, Application of high-current pulsed electron beams for modifying the surface of gas-turbine engine blades. Russ. J. Non-Ferrous Met. **57**, 256–265 (2016)
112. Y. Ivanov, S. Konovalov, V. Gromov, Electron-beam modification of the pearlite steel. Arab. J. Sci. Eng. **34**(2A), 233–243 (2009)
113. L. Diankun, G. Bo, Z. Guanglin, L. Jike, H. Liang, High-current pulsed electron treatment of hypoeutectic Al–10Si alloy. High Temp. Mater. Process. **36**, 97–100 (2017)
114. X. Cao, H. Jing, W. Huo, X. Xi, W. Zhao, Surface microstructure and property modifications in AISI 304 stainless steel induced by pseudospark pulsed electron beam treatments. Vacuum **184**, 109914 (2021). https://doi.org/10.1016/j.vacuum.2020.109914
115. D.A. Bessonov, S.V. Vorob'ev, Y.F. Ivanov, Fatigue life increase of 20Kh13 steel by electron beam processing. Izv. Vyssh. Uchebn. Zaved., Chern. Metall. **10**, 48–49 (2011)
116. D.A. Bessonov, S.V. Vorob'ev, V.E Gromov et al., Formation of nanocrystalline structures in stainless steel subjected to electron beam processing and fatigue cycling. Nanoinzheneriya **3**, 20–24 (2013)
117. Yu.F. Ivanov, V.E. Gromov, S.V. Vorob'ev, D.A. Bessonov et al., Structure and phase composition of 20Kh13 steel surface layer after high-intensity electron beam processing. Fizich. Mezomekh. **14**(6), 111–115 (2011)
118. Y.F. Ivanov, V.E. Gromov, D.A. Bessonov et al., Evolution of structure and phase composition of 20Kh13 steel during electron beam processing and siccessive fatigue loading. Deformatsiya and razrushenie materialov **12**, 19–23 (2011)
119. V.V. Sizov, V.E. Gromov, Y.F. Ivanov et al., Evolution of grain structure of 20Kh23N18 steel surface layer subjected to electron beam processing and multi-cyclic loading. Izv. Vyssh. Uchebn. Zaved., Chern. Metall. **10**, 56–60 (2012)
120. V.V. Sizov, V.E. Gromov, Y.F. Ivanov et al., Fatigue failure of stainless steel after electron beam processing. Izv. Vyssh. Uchebn. Zaved., Chern. Metall. **6**, 35–37 (2012)
121. V.V. Sizov, V.E. Gromov, Y.F. Ivanov et al., Formation and evolution of grain structure of stainless steel during electron beam processing and fatigue cycling. Fundamental'nye problemy sovremennogo materialoved. **2**, 136–140 (2012)
122. Y.F. Ivanov, V.E. Gromov, V.V. Sizov, et al., Fatigue life increase of stainless steel by electron beam processing. Problemy chernoi metallurgii i materialoved. **1**, 66–75 (2012)
123. Yu.F. Ivanov, N.N. Koval', S.V. Gorbunov et al., Multicyclic fatigue of stainless steel treated by a high-intensity electron beam: surface layer structure. Russ. Phys. J. **54**(5), 575–583 (2011)
124. V.E. Gromov, Y.F. Ivanov, S.V. Gorbunov, Modification of structure and phase composition of stainless steel surface by electron beam processing. Problemy chernoi metallurgii i materialoved. **4**, 31–37 (2010)
125. S.V. Gorbunov, S.V. Vorob'ev, Y.F. Ivanov et al., Formation of surface layer gradient structure during electron beam processing of austenitic steel. Fiz. khim obrabotki mater. **1**, 61–65 (2011)
126. Yu.F. Ivanov, S.V. Gorbunov, S.V. Vorob'ev, Surface layer structure of 08Cr18Ni10T steel after high-intensity electron beam processing and fatigue cycling. Fizich. Mezomekh. **14**(1), 75–82 (2011)
127. Yu.F. Ivanov, N.N. Koval, S.V. Gorbunov, S.V. Vorobyov, S.V. Konovalov, V.E. Gromov, Multicyclic fatigue of stainless steel treated by a high-intensity electron beam: surface layer structure. Russ. Phys. J. **54**(5), 575–583 (2011)
128. Yu.F. Ivanov, V.E. Gromov, S.V. Gorbunov et al., Gradient structural phase states formed in steel 08Kh18N10T in the course of high-cycle fatigue to failure. Phys. Met. Metallogr. **112**(1), 81–89 (2011)

129. L. Jian, Z. Song, H. Qin, H. Huang, X. Sui, Y. Weng, Z. Mo, K. Wang, X. Ren, Surface nanocrystallization and mechanical properties of mold steel induced *via* scanning electron beam treatment. Vacuum **218**, 112634 (2023). https://doi.org/10.1016/j.vacuum.2023.112634
130. H. Wang, S. Hao, Surface nanostructure and improved microhardness of $40CrNiMo_7$ steel induced by high current pulsed electron beam treatment. Nucl. Instrum. Methods Phys. Res. B **403**, 45–50 (2017). https://doi.org/10.1016/j.nimb.2017.05.002
131. F. Yulei, J. Hua, X. Shen, Y. Wang, W. Zhao, Surface hardening of 30CrMnSiA steel using continuous electron beam. Nucl. Instrum. Methods Phys. Res. B **410**, 207–214 (2017). https://doi.org/10.1016/j.nimb.2017.08.014
132. Q. Qin, R. Wang, D. Wei, K. Wang, H. Zhao, J. Zhang, H. Yan, Surface modification of 5CrNiMo steel using scanning electron beam assisted by ultrasonic vibration. Mater. Today Commun. **366**, 106497 (2023). https://doi.org/10.1016/j.mtcomm.2023.106497
133. R. Wang, H. Cui, J. Huang, H. Jiang, Effect of the continuous electron beam process treatment in the surface modification of T10 steel. Nucl. Instrum. Methods Phys. Res. B **436**, 29–34 (2018). https://doi.org/10.1016/j.nimb.2018.09.004
134. M. Ormanova, P. Petrov, D. Kovacheva, Electron beam surface treatment of tool steels. Vacuum **135**, 7–12 (2017). https://doi.org/10.1016/j.vacuum.2016.10.022
135. Y. Wei, S. Qi, Y. Wang, X. Chu, Z. Sun, J. Wang, L. Zhang, W. Jia, X. Yang, S. Liu, Microstructural evolution and tribological properties of M2 high-speed steel fabricated under various selective electron beam melting processing parameters. Tribol. Int. **187**, 108749 (2023). https://doi.org/10.1016/j.triboint.2023.108749
136. Yu.F. Ivanov, V.E. Gromov, V.A. Grishunin et al., Surface layer structure and fatigue life of rail steel irradiated by high-power electron beam. Fizich. Mezomekh. **16**(2), 47–53 (2013)
137. Y.F. Ivanov, V.E. Gromov, V.A. Grishunin et al., Electron beam processing of rail steel: Phase composition, structure, fatigue life. Voprosy materialoved. **1**(73), 20–30 (2013)
138. V.A. Grishunin, V.E. Gromov, Y.F. Ivanov et al., Fatigue life improvement of rail steel by electron beam treatment. Izv. Vyssh. Uchebn. Zaved., Chern. Metall. **2**, 51–54 (2013)
139. V.E. Gromov, Yu.F. Ivanov, V.A. Grishunin et al., Scale levels of structure, phase composition and fatigue life of rail steel after electron beam processing. Uspekhi fiziki metallov. **14**(1), 67–80 (2013)
140. S.V. Panin, A.V. Koval', G.V. Trusova, Yu.I. Pochivalov, O.V. Sizova, Influence of interface geometry and structure on plastic strain development at the mesoscale level of boronised structural steel. Fizich. Mezomekh. **3**(2), 99–115 (2000)
141. F.K. Mirzoev, V.Y. Panchenko, L.A. Shelepin, Laser control processes in solids. Phys. Usp. **39**, 1–29 (1996)
142. G.A. Bleykher, V.P. Krivobokov, A.V. Yuryeva, Thermal processes and emission of atoms from the liquid phase target surface of a magnetron sputtering system. Russ. Phys. J. **58**, 431–437 (2015)
143. A.P. Yalovets, Calculation of flows of a medium induced by high-power beams of charged particles. J. Appl. Mech. Tech. Phys. **38**, 137–150 (1997)
144. B.B. Khaimzon, V.D. Sarychev, V.E. Gromov, Temperature distribution produced by pulsed energy fluxes, with evaporation of the target. Steel Transl. **43**, 55–58 (2013)
145. E.E. Slyadnikov, I.Y. Turchanovskii, Model of electron beam processing on solid surface. Izv. Vyssh. Uchebn. Zaved., Fiz. **56**(9/2), 149–150 (2013)
146. A.A. Bugaev, V.A. Lukoshkin, V.A. Urpin, D.G. Yakovlev, Thermocapillary phenomena and surface texture formation at picosecond laser pulses. Zhurnal tekhnich. fiz. **58**(5), 908–914 (1988)
147. V.D. Sarychev, S.A. Nevskii, S.V. Konovalov, I.A. Komissarova, E.V. Cheremushkina, Thermocapillary model of formation of surface nanostructure in metals at electron beam treatment. IOP Conf. Ser. Mater. Sci. Eng. **91**, 012028 (2015)
148. A.A. Uglov, S.V. Selishchev, *Autowave Processes During Concentrated Energy Flows* (Nauka, Moscow, 1987). (in Russian)
149. A.A. Samokhin, N.N. Il'ichev, P.A. Pivovarov, A.V. Sidorin, Laser vaporisation of absorbing liquid under transparent cover. Bull. Lebedev. Phys. Inst. **43**(5), 156–159 (2016)

150. S.A. Akhmanov, V.I. Emel'yanov, N.I. Koroteev, V.N. Seminogov, Interaction of powerful laser radiation with the surfaces of semiconductors and metals: nonlinear optical effects and nonlinear optical diagnostics. Sov. Phys. Usp. **28**, 1084–1124 (1985)
151. A. Oron, S.H. Davis, S.G. Bankoff, Longscale evolution of thin liquid films. Rev. Mod. Phys. **69**, 931–980 (1997)
152. F.K. Mirzade, Wave instability of a molten metal layer formed by intense laser irradiation. Technical Phys. **50**, 993–998 (2005)
153. E.D. Eidel'man, Excitation of instability in liquid semiconductor film under heating. Zhurnal tekhnich. fiz. **64**(4), 29–45 (1994)
154. E.D. Eidel'man, Influence of the thickness of the liquid layer on the ratio of the dimensions of a convection cell. Tech. Phys. **43**, 1275–1279 (1998)
155. E.D. Eidel'man, Excitation of an electric instability by heating. Phys. Usp. **38**, 1231–1246 (1995)
156. S.A. Nevskii, Formation mechanism of micro- and nanocrystalline surface layers of titanium and aluminium alloys after electron beam processing. Fundamental'nye problemy sovremennogo materialoved. **17**(10), 385–395 (2020)
157. S. Nevskii, V. Sarychev, S. Konovalov, A. Granovskii, V. Gromov, Formation mechanism of micro- and nanocrystalline surface layers in titanium and aluminum alloys at electron beam irradiation. Metals **10**(10), 1399 (2020)
158. S.A. Nevskii, S.V. Konovalov, K.A. Osintsev, Y.F. Ivanov, A.Y. Granovskii, V.E. Gromov, The mechanism of formation of surface micro- and nanostructures in the AlCoCrFeNi high-entropy alloy during electron-beam treatment. Lett. Mater. **11**(3), 309–314 (2021)
159. S.A. Nevskii, V.D. Sarychev, S.V. Konovalov, K.A. Osintsev, Y.F. Ivanov, I.A. Panchenko, V.E. Gromov, Modeling the mechanism of micro/nanostructured surface formation in Al–Co–Cr–Fe–Ni and Co–Cr–Fe–Mn–Ni high-entropy alloys treated with a high current pulsed electron beam. Lett. Mater. **12**, 3(47), 249–254 (2022)

Chapter 2
Structure and Phase Composition of Plasma-Surfaced HSS

2.1 Plasma-Surfaced Layer Structure of the Type 1.3355 Steel

According to classical concepts, the structure formation in the plasma-surfaced layer occurs as described by Geller [1]. Carbon-depleted α-solid solution precipitates from the liquid. Next, the peritectic reaction occurs at the phase interface with the formation of γ-solid solution crystals, which separate the α-crystal core from the liquid richer in carbon. Further peritectic reaction proceeds only if carbon and alloying elements diffuse from the liquid through the γ-phase. In real conditions of rapid cooling of plasma-surfaced layers, this process is absent. And the structure retains a certain content of the α-phase, which depends on the cooling rate of the surface layer. Successive cooling provides eutectoid transformation of the α-phase with the formation of a dispersed mixture of austenite and M_6C carbide and cemented carbides.

According to [1], the higher the cooling rate, the more pronounced is the structural inhomogeneity caused by the developing peritectic transformation. After complete solidification, the structure contains grains composed of three concentric layers: (1) a core with the two-phase α-eutectoid structure, (2) a bright interlayer (peritectic transformation-induced solidification provides the formation of γ-crystals, which transform into martensite and residual austenite at rapid cooling), and (3) an external layer with two-phase eutectics of austenite and carbides, which transforms into martensite and carbides after cooling.

The plasma-surfaced layer structure is characterized by well-pronounced carbide inhomogeneity in the form of the eutectic carbide network precipitating along the grain boundaries of the solid solution.

According to SEM with X-ray microanalysis (SEM/EDS), the microstructure of the plasma-surfaced layer is a common cast structure with the grain size distribution independent of the distance to the surface (see Fig. 2.1a). This can be attributed to the low thickness of the plasma-surfaced metal and, hence, close cooling rates at the depth of layers deposited one after another. One can see bright shells in this SEM image,

A. I. Potekaev et al., *High-Speed Steel After Tempering and Electron Beam Processing*, SpringerBriefs in Applied Sciences and Technology,
https://doi.org/10.1007/978-981-95-1398-7_2

which are crystals of martensite and residual austenite generated during quenching from the γ-phase involved in the peritectic reaction. Skeleton-like primary carbides Me_6C locate inside the bright shells. These carbides reduce the steel toughness, and they should be destructed one way or another. Dark areas represent a two-phase eutectic structure, which comprises carbides, martensite, and residual austenite after solidification.

Since plasma surfacing was conducted in nitrogen, resulting carbides contained nitrogen or carbonitrides. As was shown in [2, 3], complex carbides $Fe_3(W–Mo–N–V)_3C$ appeared in the structure, and the formation of Fe_4N nitrides was possible also.

Figure 2.1a presents a typical SEM image of the plasma-surfaced layer structure in the plane parallel to the substrate surface. The layer grain structure is cellular dendritic. The grain size ranges between 4 and 13.5 μm. Grain boundaries are separated by rather thin ($\approx$ 1 μm) interlayers of the secondary phase. Figure 2.1b–d, presents EDS maps acquired for different elements. Thus, the grain volume is rich in iron, and interlayers on the grain boundaries are mostly rich in tungsten, chromium, and vanadium. Aluminum is observed in the form of rounded oxygen-rich inclusions (Fig. 2.1e). The elemental composition of the plasma-surfaced layer detected by the XRD analysis is given in Table 2.1. As expected, iron is the main chemical element here, and other elements are present in a substantially lower content.

In-depth studies of the structure of the plasma-surfaced layer show the presence of higher etchability regions in the grain volume (at the center) (see Fig. 2.2). There are acicular particles of the secondary phase with the lengthwise size of 150–730 nm.

The X-ray microanalysis of the elemental composition of the plasma-surfaced layer represents the analysis of a particular point. These points are indicated in Fig. 2.3 by red squares with numbers. The results are summarized in Table 2.2.

According to Table 2.2, bright interlayers of the secondary phase observed on the grain boundaries are rich in tungsten, carbon, vanadium, and aluminum (point 1). Dark-colored rounded inclusions are rich in tungsten, chromium, and vanadium; carbon atoms are not present (point 2). Points 3, 4 are characterized by relatively low content of aluminum and vanadium; aluminum being absent at point 4.

It can thus be concluded that the structure and elemental composition of the plasma-surfaced layer structure is described by the considerably inhomogeneous distribution of iron-alloying chemical elements that should promote the formation of the multiphase structure.

2.2 The Structure of Plasma-Surfaced 1.3355 Steel Coating After High-Temperature Tempering

The structure modification of the plasma-surfaced layer occurs after the tempering process repeated 4 times at 580 °C for 1 h and successive cooling in air. Martensite and residual austenite transform into martensite tempered at increased etchability, and the

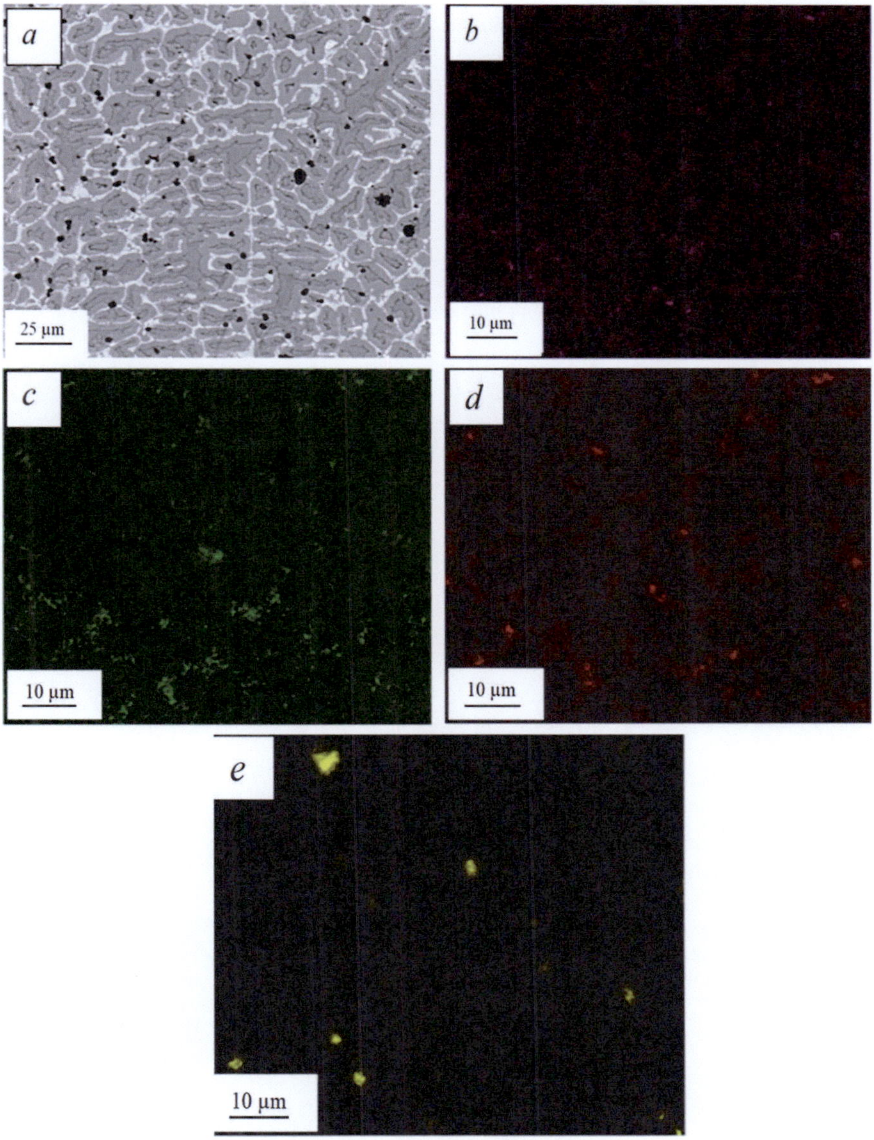

Fig. 2.1 SEM/EDS elemental analysis of plasma-surfaced layer structure: **a**—SEM image, **b**—tungsten, **c**—chromium, **d**—vanadium, **e**—aluminum distribution maps

formation of MeC, Me_6C carbides is observed. At the center of the plasma-surfaced layer, the cellular dendritic structure appears in the base metal. The metallographic analysis identifies the phase composition of the base metal as a mixture of austenite and martensite. The martensite has a well-defined acicular structure. As can be seen in Fig. 2.4a, the carbide phase in the interdendritic space has the eutectic network

Table 2.1 Elemental composition of the plasma-surfaced layer

Elements	Spectral lines	wt.%	Confidence interval, wt.%	at.%
C	K series	6.02	1.21	24.26
Al	K series	1.17	0.18	2.09
V	K series	1.29	0.11	1.22
Cr	K series	3.97	0.15	3.69
Mn	K series	0.65	0.13	0.57
Fe	K series	75.02	1.09	65.02
W	L series	11.88	0.55	3.13

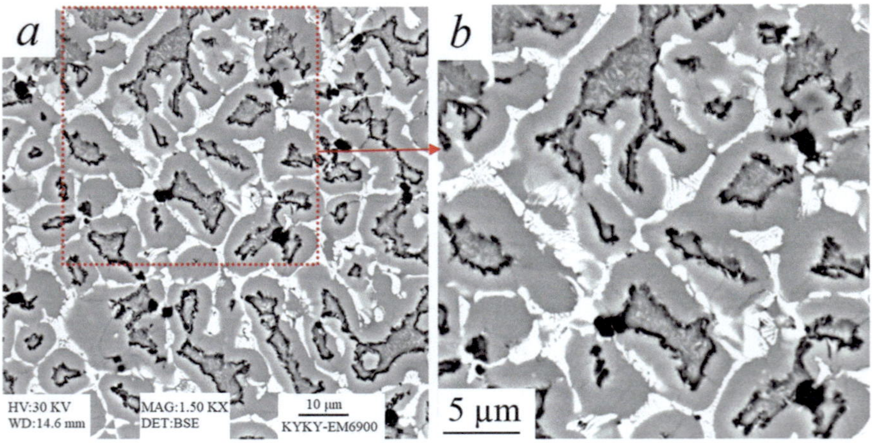

Fig. 2.2 SEM images of plasma-surfaced layer structure (**a**) and enlarged view of the fragment highlighted by red (**b**)

Fig. 2.3 SEM image of the plasma-surfaced layer structure with particular points analyzed

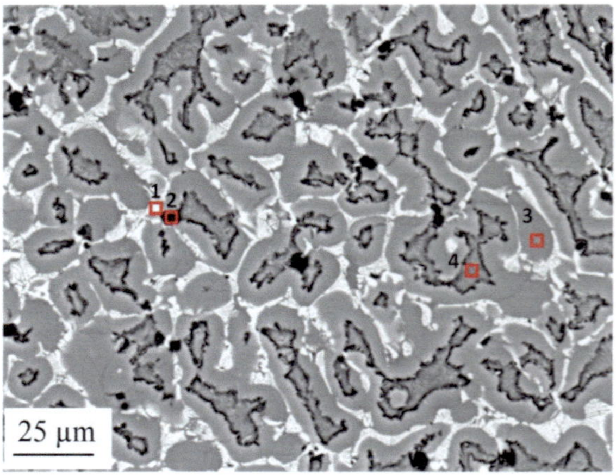

Table 2.2 Elemental composition at different points of the layer structure shown in Fig. 2.3

Elements	Point 1		Point 2		Point 3		Point 4		Average	
	wt.%	at.%	wt.%	at.%	wt.%	at.%	wt.%	at.%	wt.%	at.%
C	12.11	40.05	–	–	4.81	20.59	5.94	24.59	6.31	25.88
Al	10.08	14.83	2.99	6.78	0.55	1.04	–	–	1.10	2.01
V	7.34	5.72	4.40	5.27	1.52	1.53	1.44	1.41	2.29	2.22
Cr	3.36	2.57	4.56	5.36	4.81	4.75	3.82	3.65	4.40	4.18
Mn	0.34	0.25	–	–	0.47	0.44	0.51	0.46	0.46	0.41
Fe	44.75	31.82	70.04	76.61	73.50	67.64	74.23	66.09	69.00	60.90
W	22.01	4.75	18.0	5.98	14.34	4.01	14.06	3.80	16.44	4.41

Note The column "average" indicates the whole area of the plasma-surfaced layer

structure based on Me_6C and $Me_{23}C_6$ carbides and, probably, secondary carbide Me_2C. A confirmatory identification of the carbide phase is difficult. According to [1, 4], 10R6M5 cast alloys crystallize with the formation of Me_6C eutectic carbides, while the formation of rounded carbide particles MeC, mostly based on vanadium, occurs along the boundaries of dendritic crystals and inside them.

Figures 2.4 and 2.5 show the light microscopy images obtained at different depth of the multilayer plasma-surfaced R18Yu steel coating. One can see that the coating cross-section has the differently directed cellular dendritic structure.

The grain size distribution of this structure does not depend on the distance from the surface. This is because the thickness of individual deposited layers is low and cooling rates are thus almost the same throughout the depth of one layer. A large magnification presented in Fig. 2.5 shows no structural differences at different depth of the plasma-surfaced layer. Bright shells are martensite and residual austenite crystals formed during rapid cooling from the γ-phase involved in the peritectic reaction.

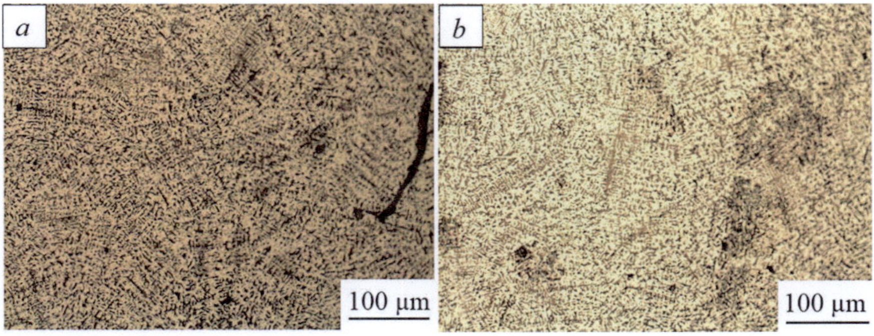

Fig. 2.4 Light microscopy cross-sectional images of plasma-surfaced R18Yu steel coating at different depths: **a**—2 mm, **b**—4 mm

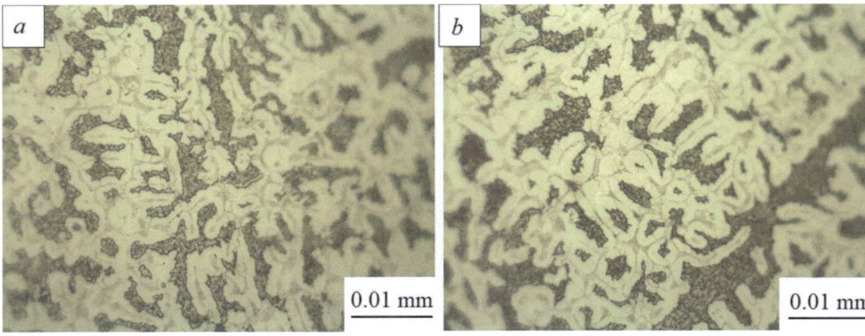

Fig. 2.5 Light microscopy cross-sectional images of plasma-surfaced R18Yu steel coating at different depths: **a**—2 mm, **b**—4 mm. Magnification: 1000×

SEM observations also do not show any difference in the layer structure at different depth. Large magnifications of this structure are presented in Fig. 2.6. After tempering (Fig. 2.7), one can clearly see the modified structure of the hard-deposited layer. Regions with martensite and residual austenite demonstrate the formation of tempered martensite with increased etchability and fine-grained carbides MeC, Me_6C.

After the tempering process repeated 4 times, decomposition of residual austenite, tempered martensite formation, and generation of fine-grained carbides resulted in the increase in the total microhardness and its more homogeneous distribution.

Successive high-temperature steel tempering is accompanied by the grain growth. The grain size ranges between 7 and 22.5 μm, as shown in Fig. 2.8a. The secondary phase interlayers locate along the grain boundaries, and their size slightly grows. Interlayers locating along the boundaries of crystallization areas are rich in W and Cr atoms, while rounded inclusions also locating here are rich in Al. V atoms distributed uniformly throughout the crystallization areas represent rounded inclusions. If we compare the results of the EDS analysis of the plasma-surfaced layer structure in

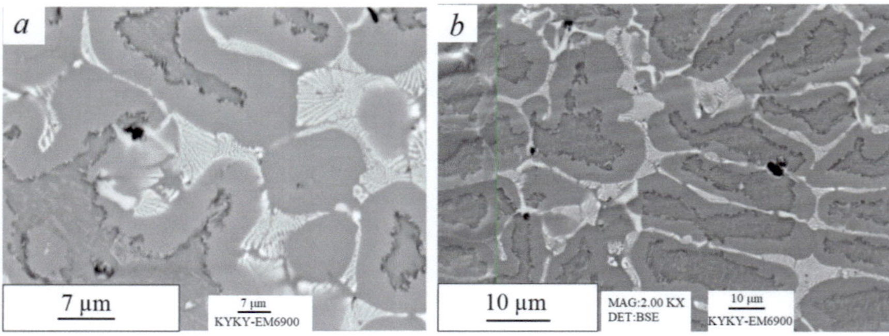

Fig. 2.6 SEM images of the layer structure at 2 mm depth at different magnifications: **a**—3500×, **b**—2000×

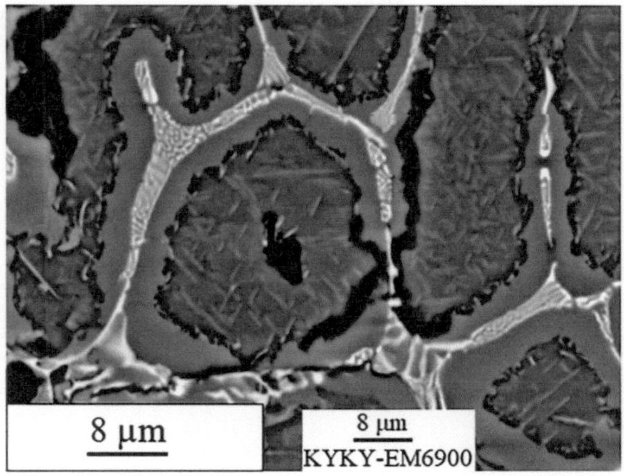

Fig. 2.7 SEM image of the layer structure at 2 mm depth after tempering. Magnification: 3000×

the initial state (Fig. 2.1b–d) and after high-temperature tempering (Fig. 2.8b–d), the latter leads to the atomic redistribution, such that Cr and V atoms move from the grain boundaries to the grain volume of the plasma-surfaced layer.

The tempering process is accompanied by the formation of lamellar (acicular) structure in the grain volume, typical for the lamellar martensite morphology. It suggests that the grain structure of the obtained layer in the initial state is provided by the γ-Fe-based solid solution (FCC austenite structure). Further tempering at high temperature leads to the redistribution of chemical elements, enabling the martensitic transformation $\gamma \rightarrow \alpha$ during the alloy cooling.

Table 2.3 presents the elemental composition of the plasma-surfaced layer after additional high-temperature tempering, detected by the X-ray microanalysis. As expected, the main element in this layer is iron, while the content of alloying elements is significantly lower.

The comparison of Tables 2.1 and 2.2 shows that high-temperature tempering has no effect on the plasma-surfaced layer. The exception is vanadium, whose content in the layer significantly reduces that can be stipulated by its nonuniform distribution in the layer structure.

The distribution of alloying elements in the tempered layer structure was analyzed at different points using EDS mapping. The results of the SEM/EDS elemental analysis are given in Fig. 2.9 and Table 2.4.

The comparison of Tables 2.2 and 2.4 shows that high-temperature tempering promotes more uniform distribution of alloying elements in the alloy structure.

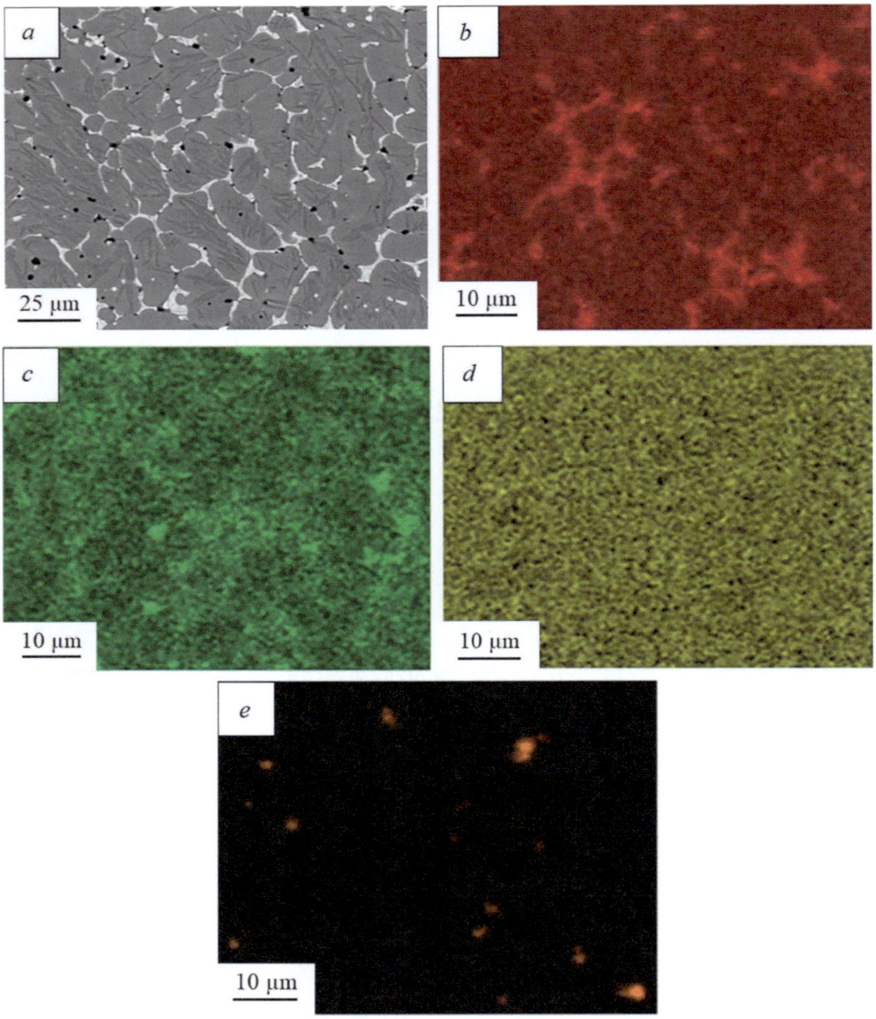

Fig. 2.8 SEM/EDS elemental analysis of plasma-surfaced layer structure after high-temperature tempering: **a**—SEM image, **b**—tungsten, **c**—chromium, **d**—vanadium, **e**—aluminum distribution maps

2.3 Plasma-Surfaced 1.3355 Steel Structure After High-Temperature Tempering and Electron Beam Processing

As stated above, Figs. 2.4 and 2.5 present the microstructure of the multilayer plasma-surfaced coating at different depths. One can clearly see the dendritic structure. First-order dendritic axes are sometimes randomly oriented relative to the surface

Table 2.3 Elemental composition of the plasma-surfaced layer after additional high-temperature tempering

Elements	Spectral lines	wt.%	σ, wt.%	at.%
C	K series	6.11	0.11	24.87
Al	K series	0.91	0.02	1.64
V	K series	0.01	0.01	0.01
Cr	K series	4.15	0.01	3.90
Mn	K series	0.36	0.01	0.32
Fe	K series	75.03	0.10	65.68
W	L series	13.44	0.05	3.57

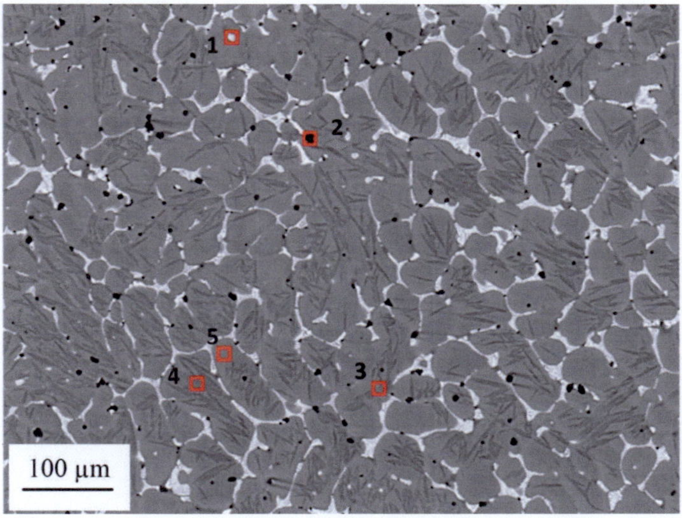

Fig. 2.9 SEM image of plasma-surfaced layer structure after high-temperature tempering with particular points analyzed

and achieve 100 μm length. The grain size distribution does not change with depth. This is because the thickness (~ 2 mm) of individual deposited layers is low and cooling rates are thus almost the same throughout the depth of one layer.

Electron beam processing provides the microcrack healing, martensite tempering, and diffusion of the carbide network. The obtained structure is shown in the SEM image with back-scattered electrons in Fig. 2.10. Electron beam processing of the deposited layer subjected to a preliminary high-temperature tempering substantially modifies its structure to a depth of 50 μm (Fig. 2.11a). The carbide network remains unchanged.

Crack healing occurs in small localized austenitic areas, not numerous transcrystalline cracks being remained. In small areas with the austenite–martensite structure,

Table 2.4 Elemental composition of plasma-surfaced layer after additional high-temperature tempering (see Fig. 2.9)

Elements	Point 1		Point 2		Point 3		Point 4		Point 5	
	wt.%	at.%	wt.%	at.%	wt.%	at.%	wt.%	at.%	wt.%	at.%
C	3.97	17.13	4.36	18.24	4.93	20.70	8.06	30.29	4.31	17.99
O	–	–	–	–	–	–	–	–	0.74	2.32
Al	0.57	1.09	2.27	4.23	0.65	1.21	0.53	0.89	0.58	1.07
Si	0.84	1.56	0.43	0.77	0.19	0.35	–	–	–	–
P	0.28	0.48	–	–	–	–	–	–	–	–
V	–	–	0.03	0.02	0.01	0.01	0.05	0.04	0.01	0.01
Cr	4.03	4.02	4.38	4.24	3.93	3.81	3.41	2.96	3.72	3.59
Mn	0.44	0.41	0.33	0.30	0.44	0.40	0.38	0.31	0.38	0.35
Fe	74.77	69.35	74.16	66.78	75.01	67.76	75.66	61.13	80.16	71.92
Ni	2.85	2.51	2.71	2.32	2.89	2.49	2.77	2.13	–	–
W	12.24	3.45	11.33	3.1	11.95	3.28	9.13	2.24	10.10	2.75

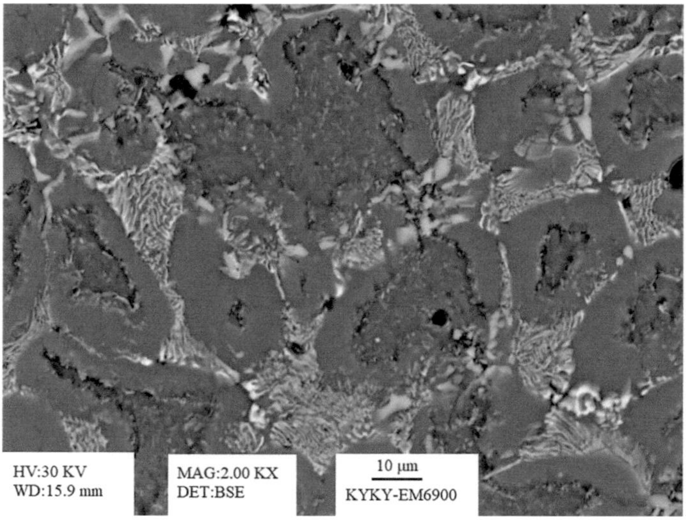

Fig. 2.10 SEM image of deposited layer structure at 2 mm depth after tempering and electron beam processing. Magnification: 2000×

acicular martensite disappears, and isolated carbides grow. The carbide network degrades.

Thermal stresses induced by rapid cooling of the material result in the surface layer fragmentation induced by microcracking, as shown in Fig. 2.11b. The cellular structure in Fig. 2.11c, appears after high-speed melting and rapid solidification of the surface layer. The cell size ranges within 0.7 and 1.0 μm. The secondary phase

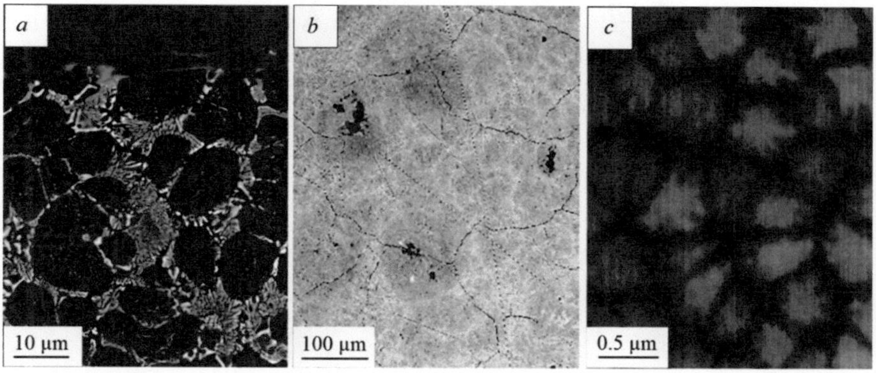

Fig. 2.11 SEM images of the layer structure after high-temperature tempering and electron beam processing: **a**—high-temperature tempering at 50 μm depth, **b**—fragmentation induced by microcracking, **c**—cellular structure

inclusions locate along the boundaries; their particle size being varied from 110 to 240 nm.

Table 2.5 presents the relative content of chemical elements forming the plasma-surfaced layer after high-temperature tempering and electron beam processing detected by the X-ray microanalysis.

According to Tables 2.1, 2.3, and 2.5, electron beam processing of the plasma-surfaced layer results in a 1.4 times reduction in the C content, 2 times increase in the V content, and 1.2 and 1.6 times increase in the content of Cr and W relative to the initial layer, respectively. It can thus be concluded that high-speed melting and successive solidification lead to a significant content growth of alloying elements and a reduction in the C content in the surface layer.

Electron beam processing of the plasma-surfaced layer makes its surface more homogeneous. The results of the X-ray microanalysis presented in Fig. 2.12 show that the atomic distribution of iron and alloying elements is quasi-uniform. This fact

Table 2.5 Elemental composition of the plasma-surfaced layer after high-temperature tempering and electron beam processing

Elements	Spectral lines	wt.%	σ, wt.%	at.%
C	K series	4.38	0.21	19.52
Al	K series	1.06	0.03	2.08
V	K series	2.35	0.02	2.43
Cr	K series	4.36	0.03	4.43
Mn	K series	0.35	0.02	0.33
Fe	K series	70.52	0.18	66.61
W	L series	16.98	0.11	4.87

distinguishes this material state from two other states discussed above (the initial plasma-surfaced layer and after high-temperature tempering).

It is expected that the surface modification of the plasma-surfaced layer after high-temperature tempering and pulsed electron beam processing leads to a change in mechanical (strength) and tribological (wear and friction coefficient) properties of the material.

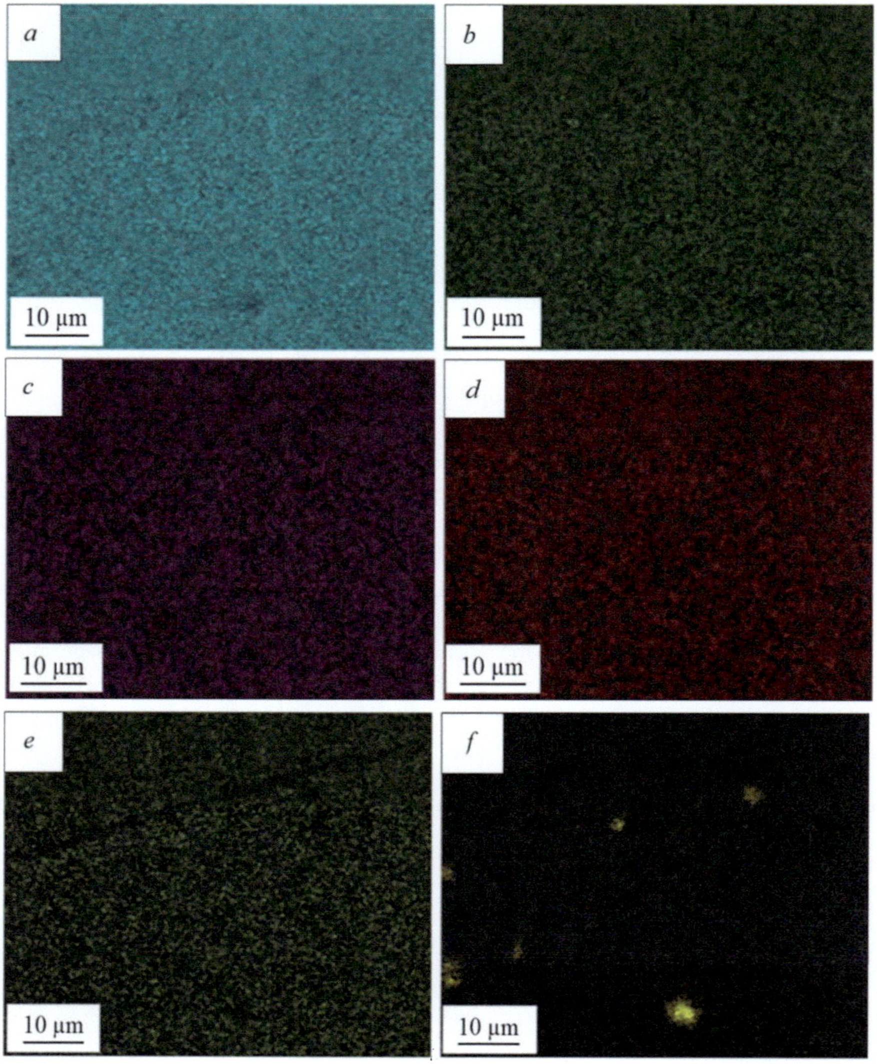

Fig. 2.12 SEM/EDS images of plasma-surfaced layer structure obtained in characteristic X-rays after high-temperature tempering and electron beam processing: **a**—iron, **b**—chromium, **c**—tungsten, **d**—vanadium, **e**—silicon, **f**—aluminum

2.4 Properties of Plasma-Surfaced Layers

Results of the microhardness measurements of the plasma-surfaced layer of the type R18Yu steel deposited onto the surface of the ČSN 14331 steel are presented in Table 2.6. One can see that the microhardness values averaged by 25 measurements are the lowest after plasma surfacing of the R18Yu steel layer in the nitrogen atmosphere (specimen 0). This parameter grows by 13% after additional tempering (specimen 1) and electron beam processing (specimen 2).

Tribology test results summarized in Table 2.6 for the plasma-surfaced R18Yu steel layer deposited onto the type ČSN 14331 steel show that high-temperature tempering provides the growth in the wear rate (decreases the wear resistance of the plasma-surfaced layer) and the reduction in the friction coefficient. The additional electron beam irradiation of tempered specimens reduces the wear rate (increases the wear resistance of the material) by 3 times and higher and the friction coefficient. Apparently, the multiple growth in the wear resistance of the plasma-surfaced layer after electron beam processing is provided by the modification of the material structure and phase composition that occurs at high-speed melting and rapid solidification of the plasma-surfaced layer, namely by the formation of submicro- and nanocrystalline structure strengthened by nanosized formations of the secondary phase.

The submicro- and nanocrystalline structure in the surface layer after the electron beam irradiation has a well-defined effect on wear tracks produced by tribology testing. Figure 2.13 presents wear track profiles for the plasma-surfaced R18Yu steel layer.

One can clearly see the rough surface of wear tracks. The roughness increases after plasma surfacing in nitrogen (Fig. 2.13a) and after plasma surfacing and high-temperature tempering (Fig. 2.13b) and reduces after electron beam processing (Fig. 2.13c). This is probably due to tungsten carbide inclusions in the plasma-surfaced layer, which spall during tribology testing. After pulsed electron beam processing (Fig. 2.13c), the roughness reduces due to the refinement of secondary-phase inclusions induced by high-speed melting followed by rapid solidification of the plasma-surfaced layer initiated by electron beam processing.

The indentation hardness H_{it} and indentation modulus E_{it} of the plasma-surfaced R18Yu layer and that one after high-temperature tempering were measured on

Table 2.6 Microhardness and tribological properties of the plasma-surfaced R18Yu steel layer

Specimens	Microhardness, GPa	Wear, $mm^3/(N \cdot m)$, 10^{-6}	Friction coefficient
0	4.7	8.9	0.70
1	5.3	10.0	0.65
2	5.3	3.0	0.58

Note Specimen 0—after R18Yu steel plasma surfacing in nitrogen, specimen 1—after R18Yu steel plasma surfacing in nitrogen and high-temperature tempering, and specimen 2—after R18Yu steel plasma surfacing in nitrogen, high-temperature tempering, and pulsed electron beam processing

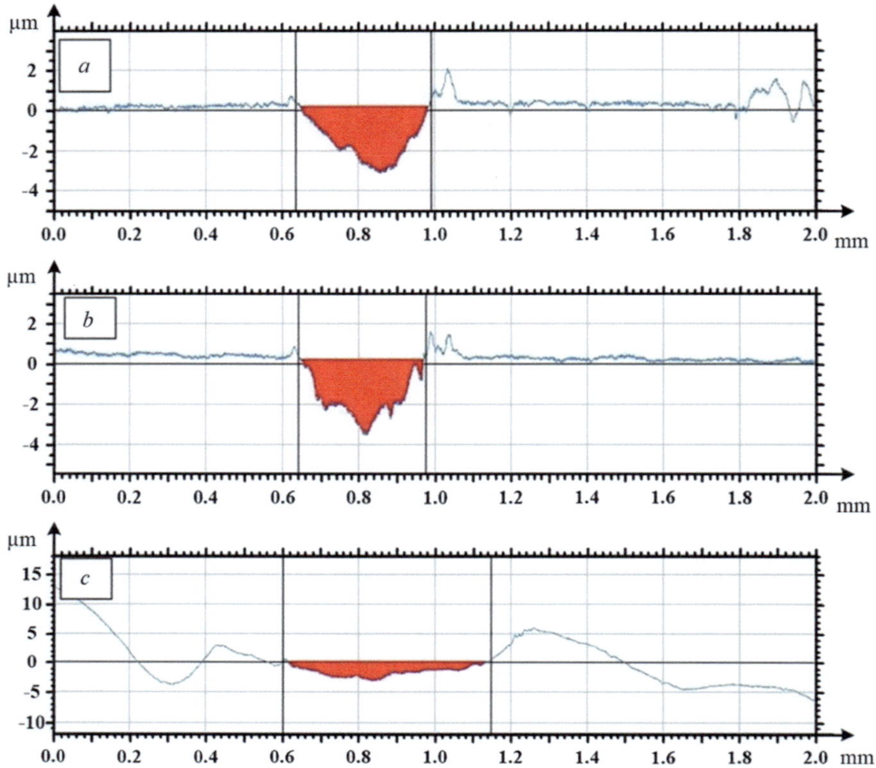

Fig. 2.13 Wear track profiles for plasma-surfaced R18Yu steel layer: **a**—after plasma surfacing in nitrogen, **b**—after plasma surfacing and high-temperature tempering, **c**—after plasma surfacing, high-temperature tempering and electron beam processing

sections of two wear tracks situated parallel to the plasma-surfaced surface. Testing conditions included 100 and 5000 μm indentation depth, 100 μm distance between indentations, and 50 mN load. Figures 2.14 and 2.15 show dependences of the hardness H_{it} and indentation modulus E_{it} on the number of indentations for the R18Yu layer after plasma surfacing in nitrogen and after high-temperature tempering, respectively.

According to Fig. 2.14, the structural inhomogeneity is high for the plasma-surfaced layer. The alloy hardness (Fig. 2.14a) changes within 4000 and 10,000 MPa. The indentation modulus also changes in a wide range. This is associated with the presence of α- and γ-Fe phases having low hardness, and carbides and carbonitrides having high hardness values.

As can be seen from Fig. 2.15, high-temperature tempering enables the formation of more uniform hardness due to the higher hardness of α-Fe grains provided by the martensitic transformation $\gamma \rightarrow \alpha$, as was indicated above.

In Fig. 2.16, H_{it} and E_{it} dependences are obtained for the plasma-surfaced R18Yu layer after high-temperature tempering and electron beam processing. Measurements

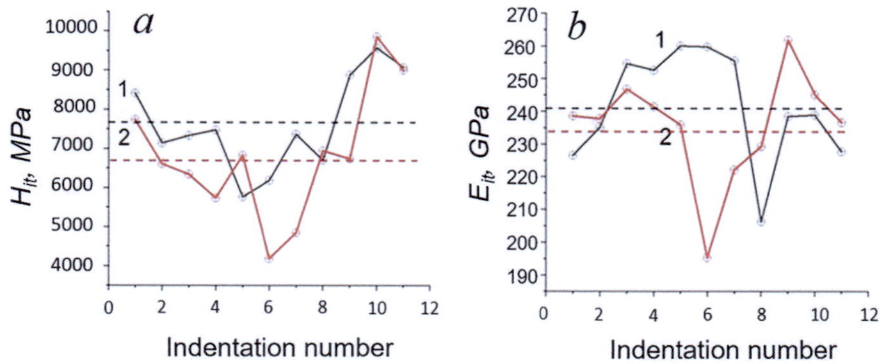

Fig. 2.14 H_{it} (**a**) and E_{it} (**b**) dependences on the number of indentations at different depths for the plasma-surfaced R18Yu layer: 1—100 μm, 2—5000 μm. Dashed lines indicate average H_{it} and E_{it} values

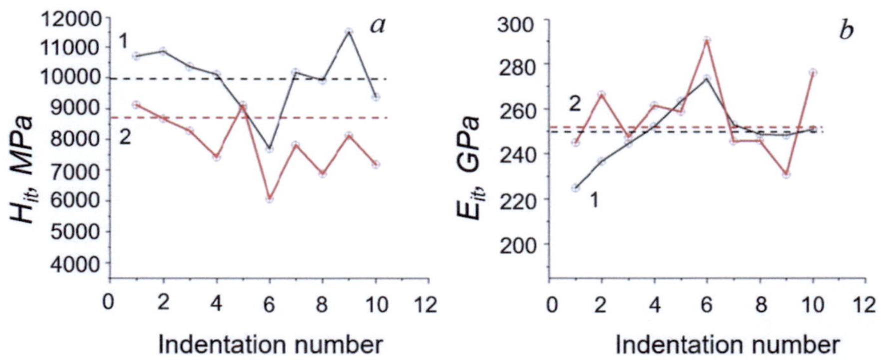

Fig. 2.15 H_{it} (**a**) and E_{it} (**b**) dependences on the number of indentations at different depths for plasma-surfaced and tempered R18Yu layer: 1—100 μm, 2—5000 μm. Dashed lines indicate average H_{it} and E_{it} values

are conducted on sections of three wear tracks (1, 2, 3) at 100 μm distance between indentations and 50 mN load.

According to Fig. 2.16, the mechanical properties of the obtained layer become considerably higher after high-temperature tempering and electron beam processing. As for the wear track 3, its results differing from the others can be explained by cracks appeared in this surface layer (see Fig. 2.11b).

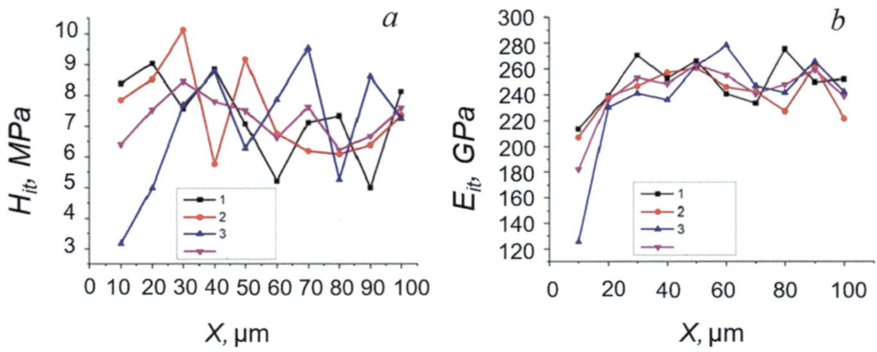

Fig. 2.16 H_{it} (**a**) and E_{it} (**b**) dependences on the distance from the surface of the plasma-surfaced and tempered R18Yu layer after electron beam processing

2.5 Conclusions

For the first time, the type 1.3355 steel coating was deposited onto the structural ČSN 14331 alloy surface using plasma surfacing in nitrogen. It was shown that successive high-temperature tempering and electron beam processing modified the structure and properties of the obtained layers. Based on the results, it was concluded that (i) plasma surfacing in nitrogen led to the layer formation of the dendritic structure with the grain size ranging between 4 and 13.5 µm. The grain volume was rich in iron. Grain boundaries were separated from each other by $\approx$ 1 µm thick secondary-phase interlayers rich in Cr, Al, W, and V atoms. The formation of small areas was detected in the austenite–martensite structure. The obtained coating has the misoriented dendritic structure with the first-order axes 100 µm long. This structure did not change at different depth. The mesh size of the carbide network detected by larger magnifications, ranged from 20 to 100 µm, and the carbide network was edged with the homogeneous metal layer. In small areas with the austenite–martensite structure, aciculae ranged in size from 1 to 3 µm; (ii) high-temperature tempering was accompanied by the grain growth up to 7–22.5 µm, Cr and V motion from the grain boundaries to the grain volume of the plasma-surfaced layer, formation of the lamellar (acicular) structure typical for the lamellar martensite morphology. The acicula size varied within 2–6 µm; (iii) electron beam processing of the tempered layer resulted in a profound modification of the layer structure to a depth of 50 µm, i.e., (1) the surface layer fragmentation induced by microcracking, (2) cellular structure formation with the cell size of 0.7–1.0 µm with the secondary-phase inclusions 110 to 240 nm in size along their boundaries, (3) a 1.4 times reduction in the C content, 2 times increase in the V content, 1.2 and 1.6 times increase in the content of Cr and W relative to the initial layer, respectively. Electron beam processing provided the microcrack healing, martensite tempering, and diffusion of the carbide network; (iv) high-temperature tempering followed by electron beam processing enhanced the microhardness of the plasma-surfaced layer by 13%, its wear resistance increased by more than 3 times, and the friction coefficient reduced by 21%.

References

1. Y.A. Geller, *Tool Steels* (Metallurgiya, Moscow, 1983). (in Russian)
2. A.N. Emelyushin, E.V. Petrochenko, S.P. Nefed'ev, Comparison of structure and properties of cast and plasma-surfaced wear-resistant materials. Liteinye protsessy **11**, 141–145 (2012)
3. S.P. Nefed'ev, A.N. Emelyushin, Nitrogen effect on structure and properties of 10R6M5 steel coatings obtained by plasma deposition. Vestnik Yugorskogo gosudarstvennogo universiteta **3**(62), 33–45 (2021)
4. M.I. Gol'dshtein, S.V. Grachev, Yu.G. Veksler, *Special Steels* (MISIS, Moscow, 1999). (in Russian)

Chapter 3
Structure and Phase Composition of Plasma-Surfaced R18Yu Steel After Tempering and Electron Beam Processing

3.1 TEM Investigation of Structure and Properties of Plasma Surfacing in Nitrogen with Nonconductive 1.3355 Cored Filler Wire on ČSN 14331 Steel

As reviewed in earlier chapters, mining, construction, and metal industries utilize plasma surfacing to protect metals and alloys from different external influences such as wear, corrosion, and static and dynamic loads and to provide them with a set of high mechanical properties [1–3]. The foremost problem is to fabricate coatings with high-performance properties, ensuring higher reliability and longer life time of products in extreme conditions (high mechanical loads, wear, corrosion environments, and thermal cycling). This problem is solved through the development and implementation of new surfacing materials containing chemical compounds, cemented carbides, and solid solutions [4]. The determinant of hardening is a selected coating material, which differs from the base material of parts and provides the desired properties for the surface.

In recent years, research and development have been intensively conducted into plasma surfacing (R18, R6M5, R2M9) using nitrogen as an alloying element [5–9]. This alloying is the best, since it allows to considerably improve the corrosion and impact resistance, strength, abrasive wear, and other properties of alloys. This becomes possible due to the improved microhardness of structural components of coatings. Reverse-polarity plasma surfacing with noncurrent carrying cored filler wire in the protective nitrogen medium has a number of advantages over other surfacing techniques [2, 3]. Rationale for the selection of the hard-facing material, meeting the operational conditions, must be proven by extensive studies of its structure, phase composition, mechanical and tribological properties, and their evolution during successive heat treatments [10–16].

A. I. Potekaev et al., *High-Speed Steel After Tempering and Electron Beam Processing*, SpringerBriefs in Applied Sciences and Technology,
https://doi.org/10.1007/978-981-95-1398-7_3

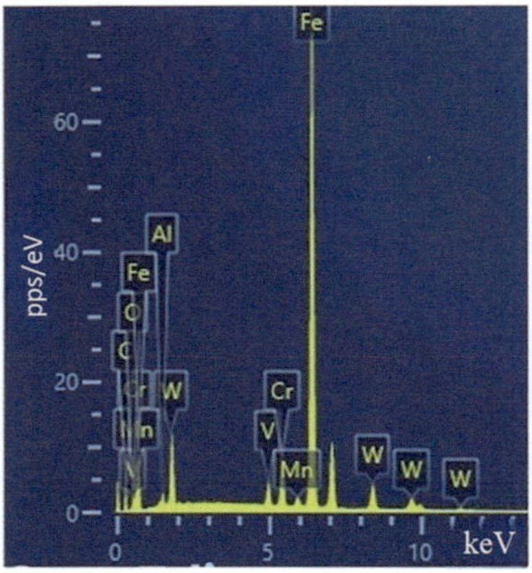

Elements	Spectral lines	wt.%	σ, wt.%	at.%
C	K series	27.21	0.15	63.48
O	K series	2.74	0.08	4.80
Al	K series	1.06	0.02	1.10
V	K series	1.62	0.01	0.89
Cr	K series	3.26	0.02	1.76
Mn	K series	0.35	0.01	0.18
Fe	K series	51.72	0.12	25.95
W	L series	12.02	0.06	1.83

Fig. 3.1 EDS spectra and atomic compositions and elemental composition of the plasma-surfaced layer

The physical nature of the formation of enhanced hardness and wear resistance of the HSS plasma surfacing can only be identified by the TEM technique [17–19]. The capacity of this technique is demonstrated in works [4, 20–22].

The aim of this chapter is TEM investigations of the structure, elemental and phase compositions, mechanical and tribological properties of the R18Yu alloy plasma surfacing onto the ČSN 14331 tool steel.

The elemental analysis of the deposited layer (see Fig. 3.1) shows the presence of all main chemical elements typical for the powder wire R18Yu. It should be noted that the plasma-surfaced layer consists of oxygen atoms that may indicate the formation of oxide phase inclusions.

Solidification of the plasma-surfaced layer formed in nitrogen on the ČSN 14331 steel surface using nonconductive R18Yu cored filler wire leads to the formation of the dendritic cellular structure, as presented in Fig. 3.2. SEM is used to characterize the morphology and microstructure of the obtained layer using secondary electrons and back-scattered electrons. Since the secondary electron (SE) signal is very sensitive to the surface topography, a secondary electron detector is used for SE imaging of the surface texture (Fig. 3.2a). The energy of secondary electrons in low (< 50 eV). Therefore, the finest obstacles encountered by secondary electrons absorb them that determines the SE sensitivity to the surface texture.

EBSD detector enables contrast images. Back-scattered electrons are a part of the primary electron beam which elastically reflects from the specimen surface. The material property to reflect electrons depends on the atomic number. Hence, BSE images visualize the difference in the average atomic number between the specimen components. As is known, the lighter the object, the higher its average atomic number. The EBSD analysis of the layer structure presented in Fig. 3.2b, shows different gray gradations in brightness. Thus, long interlayers along the grain boundaries (dendrites) are lighter that indicates to their formation by heavier elements, i.e., W-rich layers.

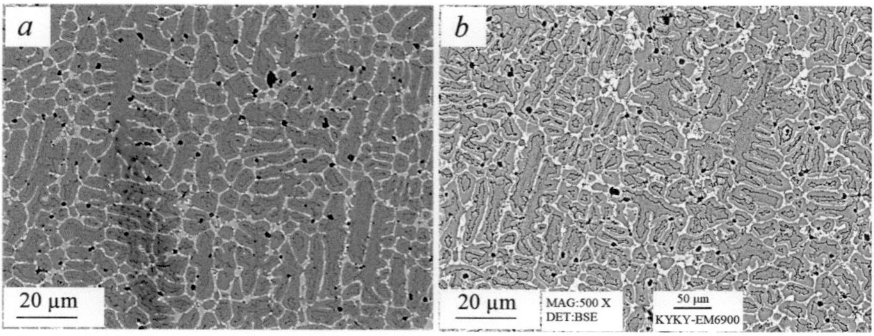

Fig. 3.2 SEM images of the R18Yu plasma-surfaced layer: **a**—obtained by SE detector, **b**—obtained by EBSD detector

Globular inclusions located chaotically in the structure are darker that means their formation by light elements, i.e., Al- and, probably, O-rich interlayers.

According to the X-ray microanalysis, secondary-phase interlayers along the grain boundaries (dendrites) are rich in heavy elements (W, V, Cr), grains are rich in lighter elements (Fe, Mn, Cr), and globular inclusions are rich in the lightest elements (Al), as shown in Fig. 3.3.

SEM observations (Figs. 3.2 and 3.4) of etched cross-sections allow us to study the material structure in detail. The structure morphology is distinguished by its elements differing in the shape, size, and mutual arrangement. Grains are the main structural element of the deposited layer, ranging in size from 7 to 22.5 μm. According to Fig. 3.4, the grain volume is divided into two contrast areas: central (darker contrast, area II) and peripheral (lighter contrast, area I). The central area is characterized by the secondary-phase inclusions.

As shown in Fig. 3.2, the obtained layer has the cellular network structure dividing it into grains (cells). Figure 3.4 shows that the cellular network has two morphological types: long and rather thin interlayers (area IV) and eutectoid structure (area III).

The XRD analysis was used to study the phase composition of the plasma-surfaced layer. It was found that the BCC α-Fe-based solid solution was the major phase (see Fig. 3.5, Table 3.1). The carbide phase $(Me)_6C$ or $(FeW)_6C$ was also clearly identified. The XRD analysis showed that in addition to the carbide phase, the layer structure could contain traces of other carbides.

As can be seen from Table 3.1, the crystal lattice parameter of the α-Fe phase is rather high. With regard to the lattice parameter $a_0 = 2.866$ Å of pure iron and carbon being the main chemical element capable of increasing the α-Fe lattice parameter, we can detect the carbon concentration ΔC in the α-Fe-based solid solution. Based on the relation proposed in [23, 24], we get $\Delta C = 0.55$ wt.%. With respect to the relative α-Fe content (0.52) in the plasma-surfaced layer, the C content is 0.286 wt.% in the α-Fe-based solid solution.

The phase and elemental compositions and defect substructure of the obtained layer were investigated on an X-ray diffractometer. Figure 3.6 and Table 3.2 present

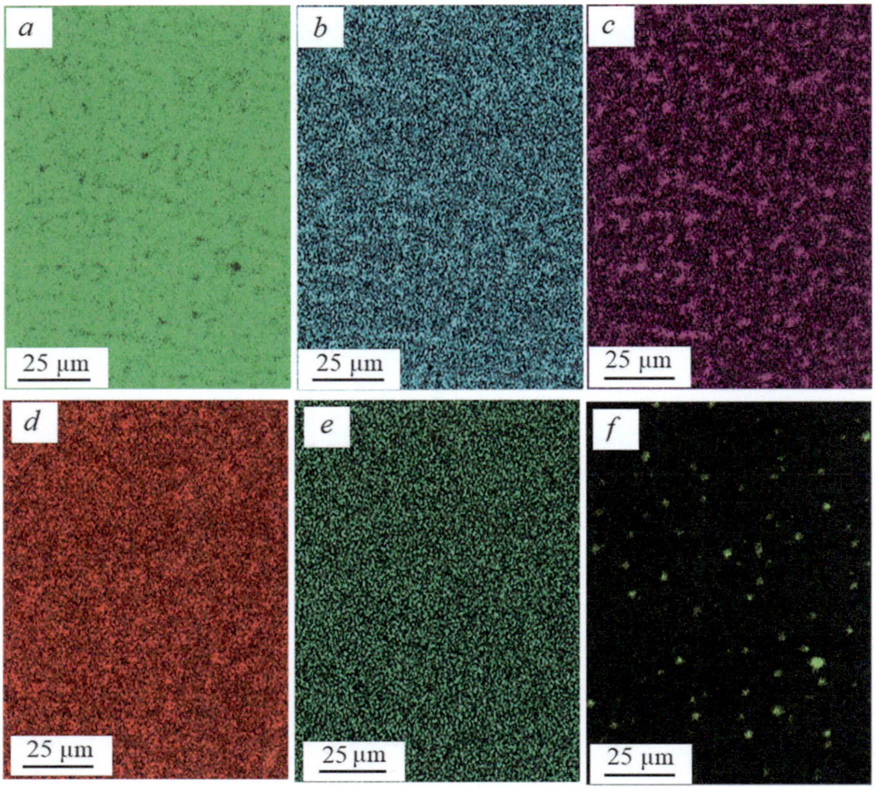

Fig. 3.3 EDS elemental mapping of plasma-surfaced layer structure obtained in characteristic X-rays: **a**—iron, **b**—tungsten, **c**—vanadium, **d**—chromium, **e**—manganese, **f**—aluminum

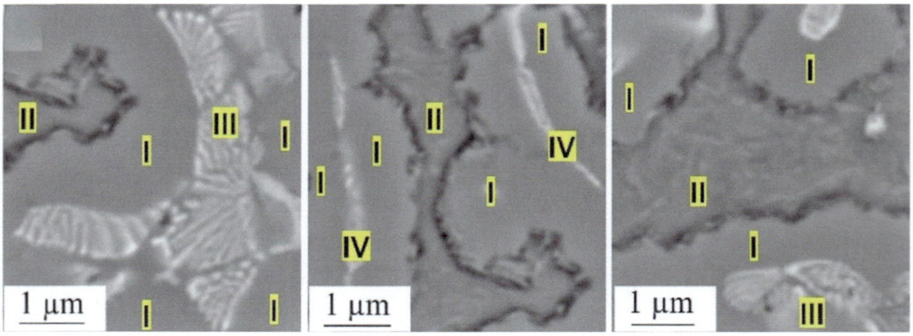

Fig. 3.4 Cross-sectional SEM images of plasma-surfaced layer with indicated contrast areas and morphology

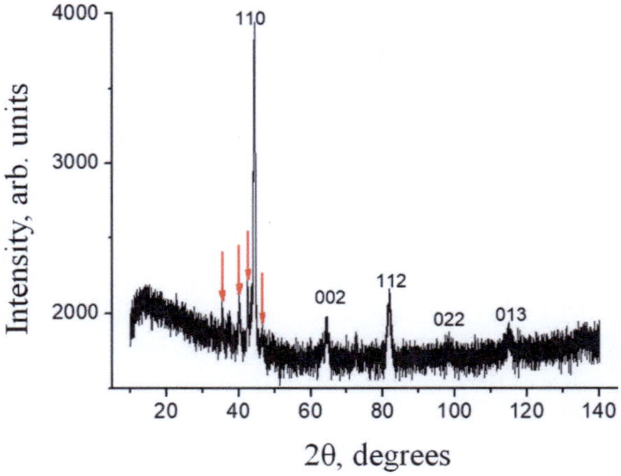

Fig. 3.5 XRD pattern for the plasma-surfaced layer. Arrows indicate diffraction peaks for Fe_3W_3C carbide

Table 3.1 Calculated parameters based on the XRD analysis (Fig. 3.5)

Phases	Volume fraction, wt.%	Lattice parameters, Å	CSR size, nm
Fe_3W_3C	48	$a = 11.096$	65
α-Fe	52	$a = 2.888$	44

these results for the carbide network structure. One can see that C, V, Cr, Fe, and W are the main elements intensively involved in the carbide phase formation.

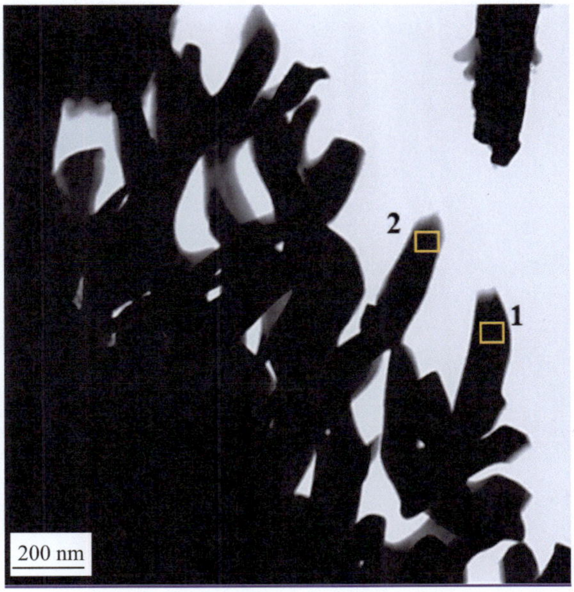

Fig. 3.6 TEM image of the plasma-surfaced layer structure. Squares indicate the analyzed areas

Table 3.2 Elemental composition of carbide network fragments

Spectra	Chemical elements, wt.%						
	C	Al	V	Cr	Mn	Fe	W
1	3.56	0.14	3.93	3.77	0.15	25.18	63.28
2	5.75	– 0.03	3.52	4.06	0.28	23.98	62.43

According to TEM images obtained for the carbide network, diffraction peaks of M_6C (Fe_3W_3C–Fe_4W_2C) carbides are observed more often in contrast to $M_{12}C$ and $M_{23}C_6$ carbides. In Fig. 3.7, TEM images and selected area diffraction (SAED) pattern of the carbide network structure demonstrate the presence of M_6C (Fe_3W_3C–Fe_4W_2C) carbides. Note that inclusions of the carbide phase do not contain the dislocation structure and are characterized by bend extinction contours, indicating to elastic stresses in the layer deposited.

The second structural element of the plasma-surfaced layer is the α-phase grains. The bright-field image given in Fig. 3.8 for their defect substructure shows the presence of the dislocation structure inside. There are grains with chaotic dislocations (Fig. 3.8a). In this case, scalar dislocation density $\langle\rho\rangle = 2.2 \times 10^{10}$ cm^{-2}. In Fig. 3.8b, the network dislocation is present in the grain, and scalar dislocation density is much higher, viz. $\langle\rho\rangle = 1.2 \times 10^{11}$ cm^{-2}.

The TEM image in Fig. 3.9 and Table 3.3 present the analysis of different areas in the α-phase grains. One can see that the elemental composition of the grain significantly depends on the material volume to be analyzed and is determined by the secondary-phase inclusions.

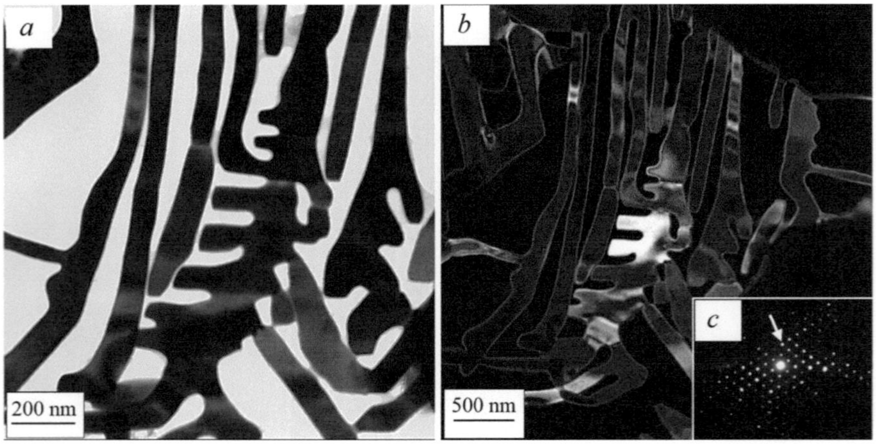

Fig. 3.7 TEM images of the carbide network structure: **a**—bright-field image, **b**—dark-field image in the [004] diffraction reflection of M_6C (Fe_3W_3C), **c**—SAED pattern

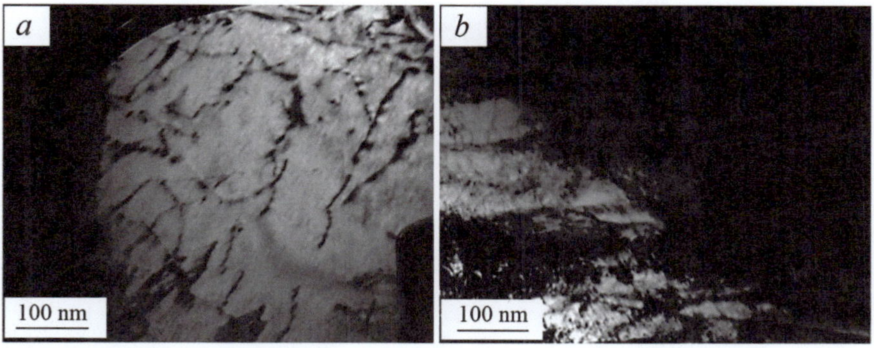

Fig. 3.8 TEM images of dislocation substructure in α-phase grains: **a**—chaotic dislocations, **b**—network dislocation

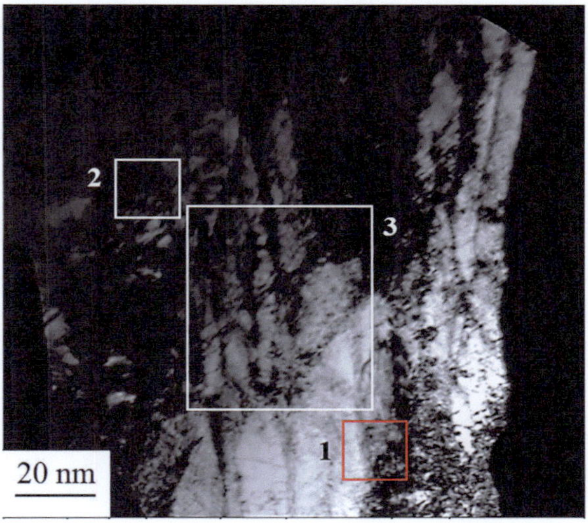

Fig. 3.9 TEM image of the plasma-surfaced layer structure. Squares indicate the analyzed areas

Table 3.3 Elemental composition of the α-phase grain

Spectra	Chemical elements, wt.%						
	C	Al	V	Cr	Mn	Fe	W
1	– 0.22	0.15	0.35	1.94	0.19	24.69	72.91
2	1.23	0.30	1.98	3.74	0.29	41.69	50.77
3	0.52	– 0.04	0.98	3.24	– 0.09	42.61	51.31

Figure 3.10 presents TEM images and SAED patterns of the secondary-phase inclusions generating in α-phase grains during the layer deposition. Bright and dark-field TEM images show the presence of the secondary-phase acicular particles. According to the SAED pattern, these particles belong to V_4C_3 carbide.

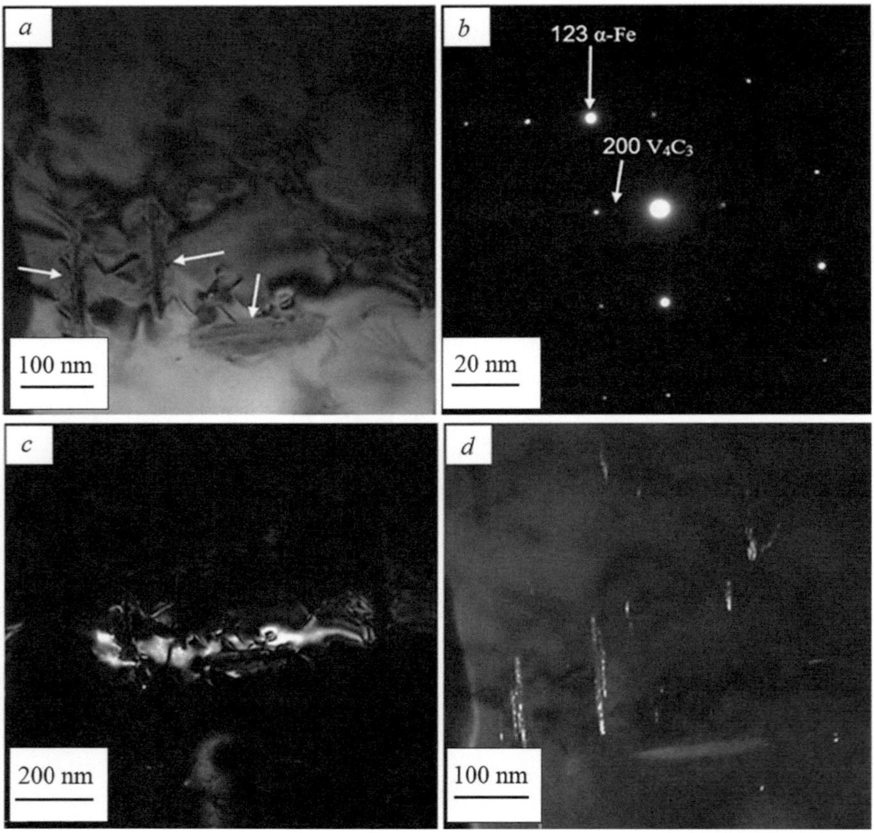

Fig. 3.10 TEM images of the α-phase grain: **a**—bright-field image, **b**—SAED pattern, **c**, **d**—dark-field images obtained in the [123] and [004] diffraction reflections of α-Fe and V_4C_3, respectively. Arrows indicate carbide phase particles (**a**) and SAED patterns for dark-field images (**b**)

Investigations of strength properties of the plasma-surfaced layer show its 4.7 GPa surface microhardness. According to tribology tests, the wear rate and friction coefficient of this layer are 8.9×10^{-6} mm^3/N · m and 0.7, respectively. Based on the results of the elemental and phase analyses, structure and properties of plasma surfacing in nitrogen with nonconductive 1.3355 cored filler wire on the ČSN 14331 steel, the following conclusions can be drawn.

- Plasma surfacing provided the formation of the layer consisting of the BCC α-Fe-based solid solution and M_6C (Fe_3W_3C–Fe_4W_2C) carbide as major phases;
- M_6C (Fe_3W_3C–Fe_4W_2C) carbide generated the carbide network and was the major strengthening phase of the plasma-surfaced layer;
- the formation of the plasma-surfaced layer was accompanied by decomposition of the α-Fe-based solid solution with the formation of nanosized carbide particles in the grain volume;

- the layer formed in nitrogen on the ČSN 14331 steel surface using nonconductive R18Yu cored filler wire was characterized by 4.7 GPa microhardness, 8.9×10^{-6} $mm^3/N \cdot m$ wear rate, and 0.7 friction coefficient.

3.2 Structure and Properties of Plasma-Surfaced Layer After High-Temperature Tempering

This paragraph focuses on the structure, elemental and phase compositions, and mechanical and tribological properties of 1.3355 plasma surfacing onto ČSN 14331 steel and subjected to the additional high-temperature tempering. The material and methods do not differ from those described in Chap. 1.

As mentioned in Chap. 2, crystallization of the plasma-surfaced R18Yu layer on the ČSN 14331 steel surface generates the structure morphology of several types, differing by shape, size, and location (Fig. 3.11a). The carbide network shown in Fig. 3.11b, divides the hard-facing material into grains. The grain size ranges between 7 and 22.5 μm. The grain volume has two regions. The central region (darker contrast) has submicro- and nanosized inclusions. The peripheral region (uniform contrast) separates the carbide network from the central region. This indicates to the presence of nanosized inclusions of the secondary phase not found after etching.

The EDS analysis of the plasma-surfaced tempered layer presented in Fig. 3.12 shows the presence of Fe, W, Cr, V, Si, Al, Mn, and C elements in its structure. The main element is iron (see Table 3.4). W, Cr, and C elements are present in much lower content. It should be noted that N is not found but O is presented that indicates to oxidation processes during surfacing.

EDS elemental mapping of chemical elements is shown in Fig. 3.13. The comparison of these EDS maps with the SEM image in Fig. 3.12b, shows that the carbide

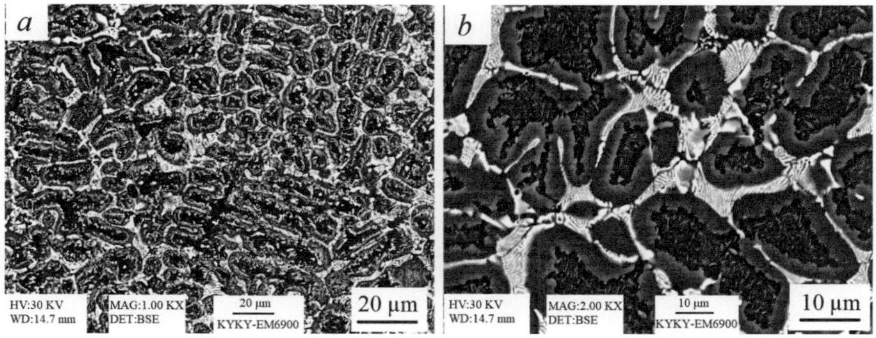

Fig. 3.11 SEM images of R18Yu plasma-surfaced structure after high-temperature tempering: **a**—morphology, **b**—carbide network

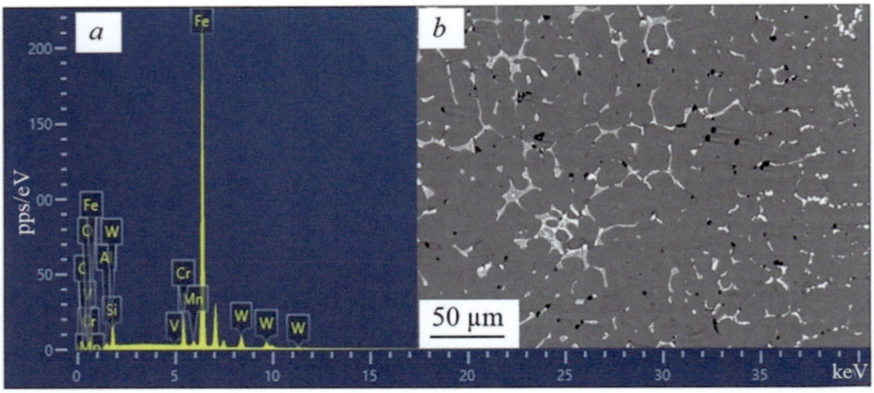

Fig. 3.12 EDS spectra (**a**) and SEM image (**b**) of the plasma-surfaced layer

Table 3.4 Elemental composition of the tempered plasma-surfaced layer

Elements	Spectral lines	wt.%	σ, wt.%	at.%
C	K series	6.25	0.15	24.51
O	K series	1.40	0.06	4.11
Al	K series	0.92	0.02	1.61
V	K series	0.02	0.01	0.02
Cr	K series	4.12	0.02	3.73
Mn	K series	0.36	0.01	0.31
Fe	K series	73.99	0.14	62.40
W	L series	12.95	0.07	3.32

network is mostly formed by W, Cr, and, probably, C atoms. Al atoms generate globular inclusions arranging chaotically in the plasma-surfaced layer and have the black contrast.

Phase and elemental compositions and the defect substructure are analyzed by energy dispersive spectroscopy (see Fig. 3.14). The EDS maps are presented for the different points of the carbide network. It is clear that its grains are rich mostly in tungsten, iron, and chromium. The presence of C atoms indicates to a probable formation of the carbide phase.

Quantification of the elemental composition is presented in Fig. 3.15 for the given fragment of the carbide network.

According to these results, tungsten and iron are the main elements of the carbide network. Chromium and vanadium are found in a much less content. It is thus suggested that the carbide network is formed by the carbide phase, which is based on tungsten and iron.

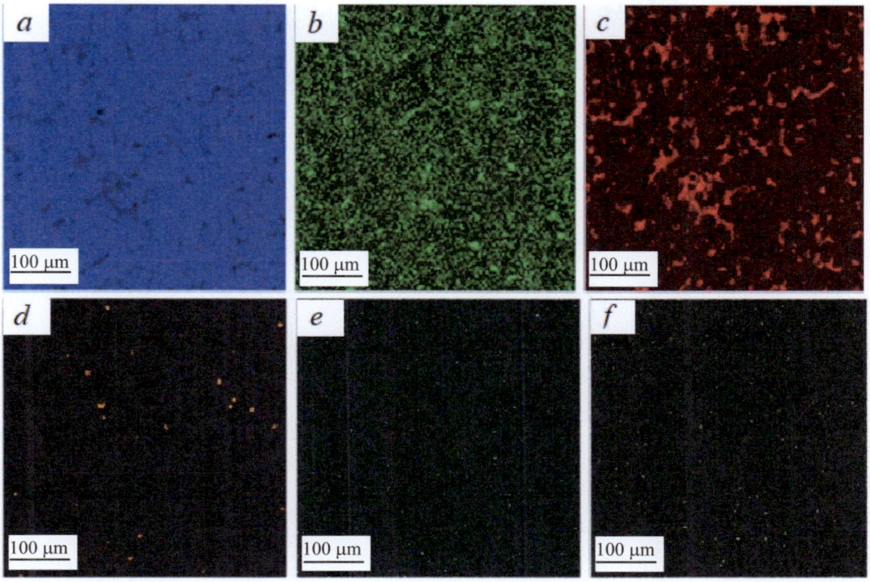

Fig. 3.13 EDS elemental mapping of plasma-surfaced layer structure obtained in characteristic X-rays for the area shown in Fig. 3.12b: **a**—iron, **b**—chromium, **c**—tungsten, **d**—aluminum, **e**—oxygen, **f**—carbon

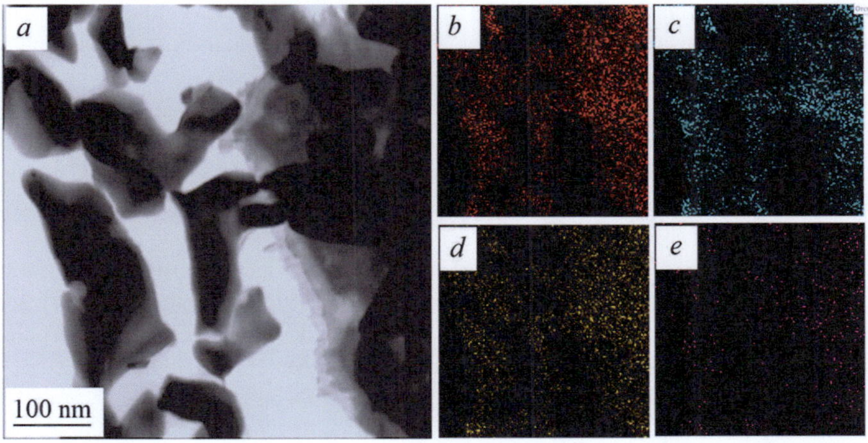

Fig. 3.14 TEM image of carbide network: **a**—bright-field image, **b**–**e**—EDS elemental mapping obtained in characteristic X-rays: **b**—W, **c**—Fe, **d**—Cr, **e**—C atoms

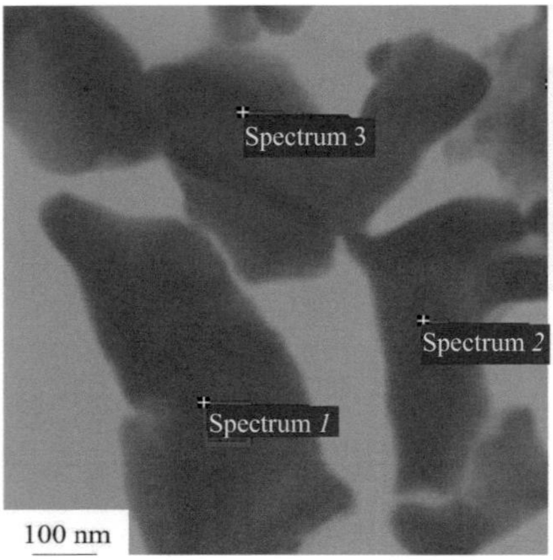

Elements, wt.%	Spectrum *1*	Spectrum *2*	Spectrum*3*
C	1.75	1.26	1.17
Al	0.17	0.37	0.0
V	2.77	3.11	2.51
Cr	3.91	4.87	4.79
Mn	0.85	0.8	0.63
Fe	21.95	19.45	21.48
W	68.78	70.14	69.42

Fig. 3.15 TEM image of the carbide network in the plasma-surfaced layer. Highlighted regions are described in the table

The SAED pattern in Fig. 3.16b obtained from the fragment of the carbide network (Fig. 3.16a) indicates that the carbide network is generated by complex carbide Fe_3W_3C (Fe_2W_4C). The grain size of the carbide phase ranges from 80 to 350 nm.

In Fig. 3.17, the TEM image and EDS elemental mapping shows the grain volume of the α-Fe-based solid solution. The presence of the carbide network inside grains indicates the formation of the carbide phase. The TEM image shows no preferred atomic arrangement for chemical elements that could indicate the elemental composition of particles, because their size is small (10–20 nm).

Quantification of the elemental composition is presented in Fig. 3.18 for the grain structure of the α-Fe-based solid solution. It is clear that the particles having the elemental composition given in Fig. 3.18 are rich in W, Cr, V, Fe, and also C atoms. The latter suggests the presence of nanosized particles of a complex carbide phase in the α-Fe-based solid solution.

Figures 3.19 and 3.20 contain TEM images and SAED patterns for the phase composition of grains in the plasma-surfaced layer structure. One can clearly see that the grain volume consists of rounded (Fig. 3.19) or faceted (Fig. 3.20) particles of the secondary phase. The particle size ranges between 10 and 38 nm.

According to these TEM observations and SAED patterns, globular particles chaotically arranged in α-Fe grains are either V_4C_3 or Cr_7C_3 carbides (Fig. 3.19). Faceted particles are either $Cr_{23}C_6$ ($M_{23}C_6$) or Fe_3C or WC_{1-x} carbides (Fig. 3.20). It is not possible to better identify these compositions due to a close position of reflections.

According to microhardness measurements of the R18Yu layer plasma surfacing onto the ČSN 14331 steel, its value averaged by 25 measurements increases by 13% relative to the nontempered state and reaches 5.3 GPa.

Tribology tests of the plasma-surfaced layer show that high-temperature tempering leads to an increase in the wear rate, i.e., the wear resistance is 12.3% lower

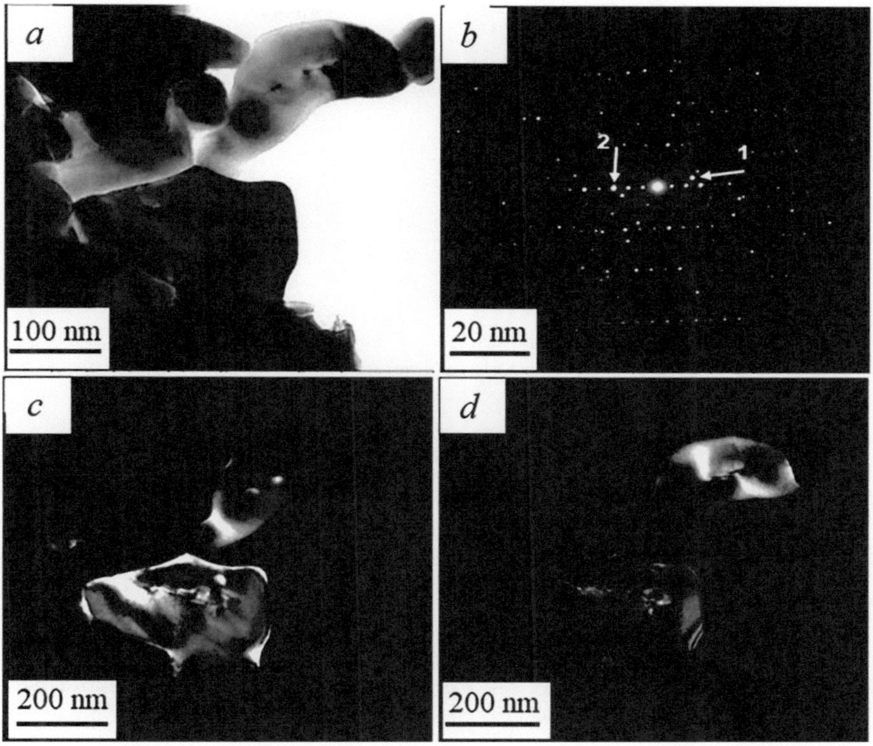

Fig. 3.16 TEM images of the carbide network of the plasma-surfaced layer: **a**—bright-field image, **b**—SAED pattern, **c**—dark-field image in the [400] diffraction reflection of Fe_3W_3C (arrow 1), **d**—dark-field image in the [422] diffraction reflection of Fe_3W_3C (arrow 2)

than that of nontempered specimen, and a 7.7% decrease in the friction coefficient relative to the nontempered state.

The indentation hardness H_{it} and indentation modulus E_{it} of the plasma-surfaced tempered layer are measured at the indentation load of 50 mN. These measurements result in $H_{it} = 10.6$ GPa and $E_{it} = 225$ GPa.

Based on the research findings concerning the structure, elemental and phase compositions, and mechanical and tribological properties of the 1.3355 steel plasma surfacing onto ČSN 14331 steel and subjected to the additional high-temperature tempering, the following conclusions can be made. The plasma-surfaced layer structure has inhomogeneous morphology and represents the carbide network. The latter is rich mostly in tungsten, chromium, and carbon and based on M_6C (Fe_3W_3C) carbide. The grain size of the carbide network ranges from 80 to 350 nm. Carbide phase particles detected inside α-Fe phase grains are either globular or faceted and range from 12 to 38 nm. Globular particles are chaotically distributed in α-Fe phase grains and represent either V_4C_3 or Cr_7C_3 carbide. And faceted particles are Fe_3C, $Cr_{23}C_6$, or WC_{1-x} carbides. It is demonstrated that high-temperature tempering is

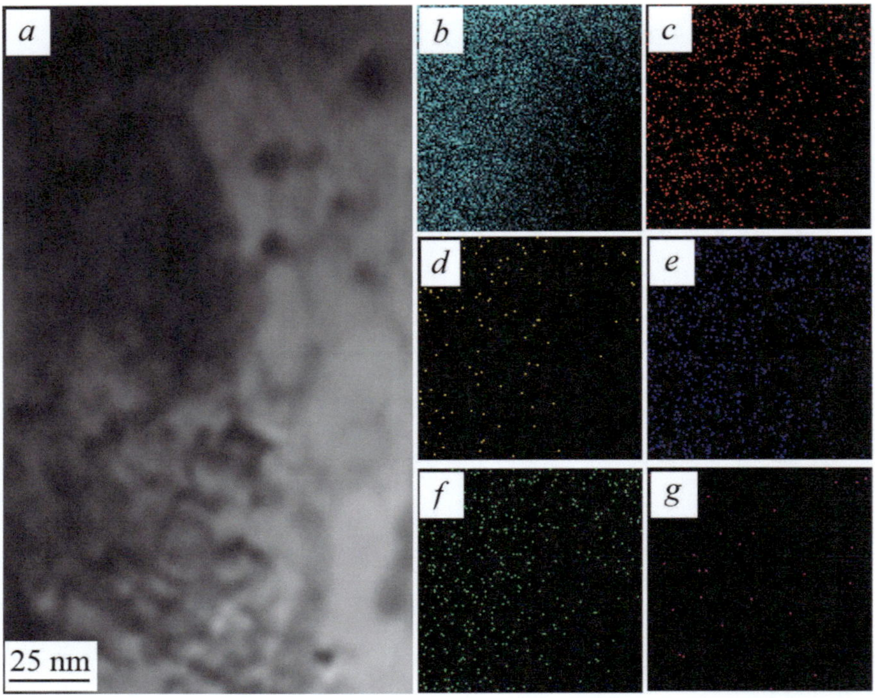

Fig. 3.17 TEM image of the grain structure of α-Fe-based solid solution: **a**—bright-field image, **b–g**—EDS elemental mapping obtained in characteristic X-rays: **b**—Fe, **c**—W, **d**—Cr, **e**—V, **f**—Mn, **g**—C atoms

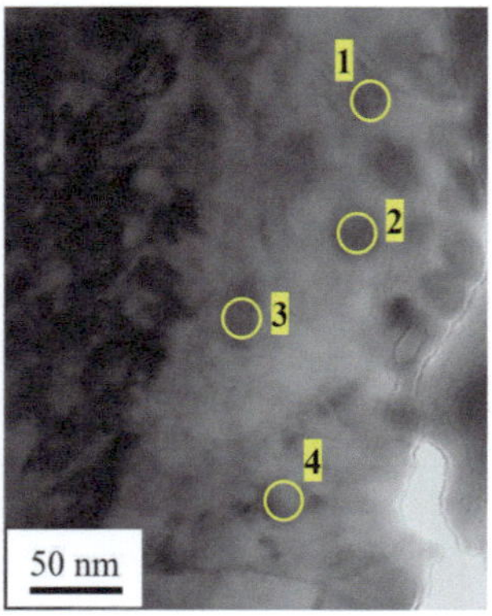

Elements, wt.%	Spectrum *1*	Spectrum *2*	Spectrum *3*	Spectrum *4*
C	0.09	0.1	0.03	0.0
Al	0.16	0.23	0.2	0.23
Si	0.2	0.3	0.14	0.39
V	6.71	4.01	2.95	4.41
Cr	4.83	4.92	4.86	6.2
Mn	0.22	0.5	0.52	0.19
Fe	41.62	55.61	60.25	45.64
W	46.17	34.33	31.05	42.94

Fig. 3.18 TEM image of the grain structure of α-Fe-based solid solution. Highlighted regions are described in the table

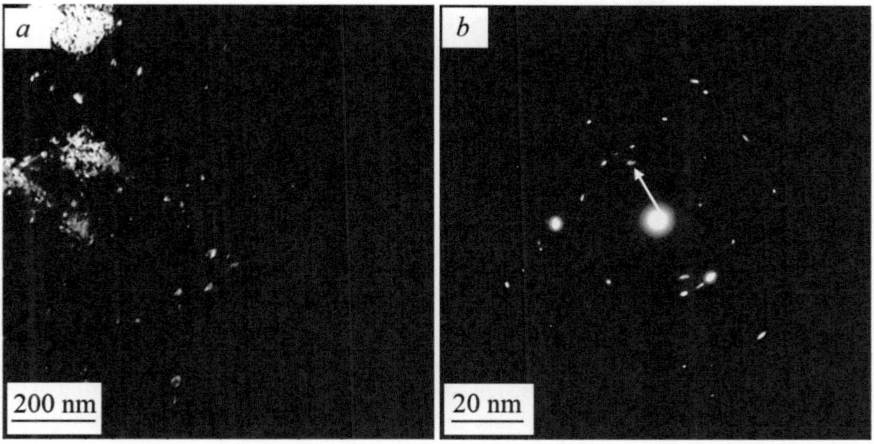

Fig. 3.19 TEM image of carbide phase particles in α-Fe phase grains: **a**—dark-field image in [110] α-Fe + [200] V_4C_3 + [421] Cr_7C_3 diffraction reflections, **b**—SAED pattern for the dark-field image

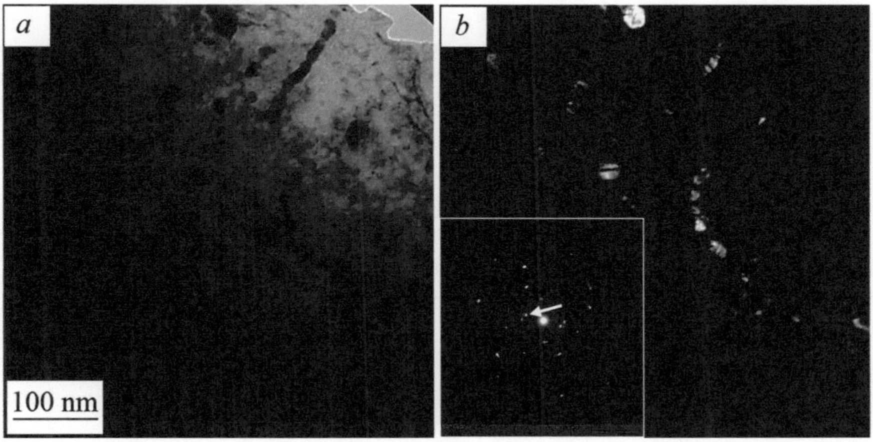

Fig. 3.20 TEM image of carbide phase particles in α-Fe phase grains: **a**—bright-field image, **b**—dark-field image in [420] $Cr_{23}C_6$ + [111] WC_{1-x} + [121] Fe_3C diffraction reflections. Insert: SAED pattern for the dark-field image

accompanied by a 13% microhardness growth of the plasma-surfaced R18Yu layer and reaches 5.3 GPa. The process is also accompanied by, respectively, 12.3 and 7.7% decrease in the wear resistance and friction coefficient of the obtained layer relative to the nontempered state.

3.3 Structure and Properties of Plasma-Surfaced Layer After Electron Beam Processing

An additional hardening of surface layers can be achieved by electron beam processing, which is one of the promising techniques of the metal surface modification, leading to a significant enhancement of the mechanical properties of the whole material. The performance of electron beam-processed metals and alloys can be increased by 20 times that is considerably higher than the performance of metals and alloys subjected to conventional treatment techniques [25–30]. High density electron beams induce different phenomena in the surface layer during a very short period of time, namely rapid recrystallization, surface polishing, and annealing. Moreover, electron beam processing can induce plastic strain of the surface layer, resulting in the formation of high-density dislocations and significant improvement of its properties.

This paragraph focuses on the structure and properties of the plasma-surfaced 1.3355 layer deposited onto a ČSN 14331 cast roll in the shielding gas (nitrogen) atmosphere, after high-temperature tempering and electron beam processing.

The SOLO installation [25] was used for pulsed electron beam processing in high purity (99.999%) argon at a pressure of 35 mPa. Using niobium foils, the specimens were fixed in the vacuum chamber, on a 5 mm distance above the water-cooled table. The electron beam parameters included 30 J/cm^2 energy density, 50 μs pulse duration (number of pulses: 5), and 0.3 s^{-1} pulse frequency. The energy density and distribution over the specimen surface were measured with a calorimeter. The discharge current of the plasma cathode and accelerating gap current were measured by Rogowski coils and recorded by Keysight DSO1014A or Rigol DS1054Z oscilloscopes (China) together with the signal of accelerating voltage from a high-voltage probe ACA-6039 from Aktakom (Russia). A Kleiber KGA 740-LO infrared pyrometer (Kleiber Infrared GmbH, Germany) was used for temperature measurements at a 6 μs response time. The Kleiber KGA 740-LO operated within 2–2.2 μm radiation band, focusing to an ~ 4 mm diameter spot. The temperature measurements were conducted in the range of 300–2300 °C. The surface emittance was detected through a comparison of the pyrometer and chromel–alumel thermocouple readouts in measuring temperature of the heated specimen. The thermocouple voltage was measured by a Fluke 175 multimeter (Fluke Corporation, USA).

Prior to electron beam processing, the specimen was sparingly heated up to 200–300 °C with the electron beam without the structure and surface modification. That improved the stability of processing parameters and prevented breakdown of the high-voltage accelerating gap during the irradiation. The specimen volume temperature measured by the chromel–alumel thermocouple, increased from initial 200–300 °C up to 390–480 °C before the 5th pulse. After the processing, specimens were cooled in vacuum until room temperature.

Temperature dependences obtained for the specimen surface during electron beam processing are presented in Fig. 3.21. One can see two regions of different temperature growth. In the first region (till the 20th microsecond), the temperature growth rate is 7.4–9.5 × 10^7 K/s. In the second region (from the 20th to the 50–60th microsecond),

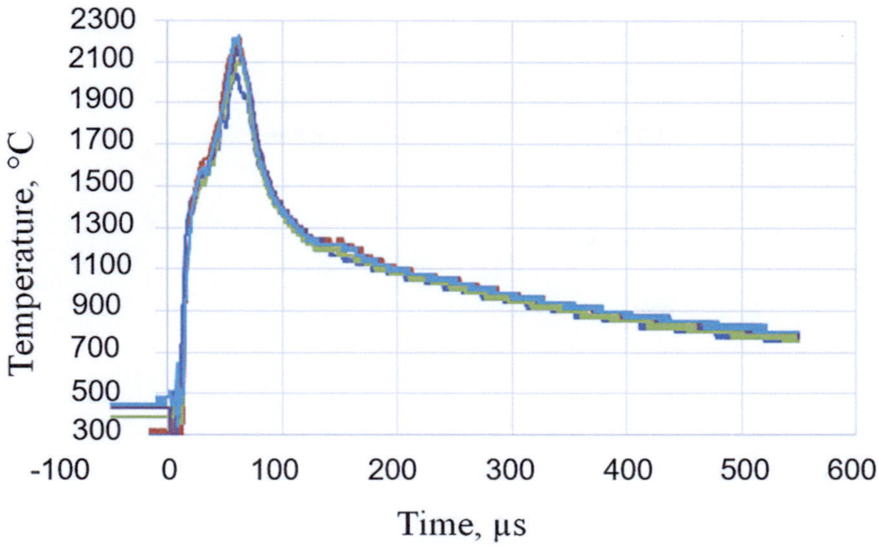

Fig. 3.21 Time/temperature dependences of the specimen surface during electron beam processing at 30 J/cm^2 energy density and 50 μs pulse duration

it is 1.2–2.2 × 10^7 K/s. After the pulse end (at 2100 °C surface temperature), the temperature drops. This region also has two sections, namely rapid (~ 1.5 × 10^7 K/s) temperature drop to the crystallization temperature (constant temperature of ~ 1200 °C for 25 μs) and temperature decrease during cooling after the layer solidification. The plasma-surfaced 1.3355 layer after the tempering process repeated 4 times at 580 °C acquires the polycrystalline structure with the grain size ranging between 7 and 22.5 μm. Interlayers of the secondary phase locate along the grain boundaries.

SEM images in Fig. 3.22 belong to the layer structure after 560 °C annealing and electron beam processing. As can be seen from Fig. 3.22b, the structure after high-temperature tempering and electron beam processing substantially modifies. First, the surface fragmentation is induced by microcracking that can indicate to relaxation of thermal stresses forming in the material after high-speed cooling. Second, the irradiation used in this experiment does not lead to a complete surface homogenization due to superhigh heating and cooling rates. There are chaotic inclusions of the secondary phase; their size ranges within units of micrometers.

The cross-sectional SEM image in Fig. 3.23a, belongs to the plasma-surfaced layer. After electron beam processing, its thickness ranges between 30 and 50 μm.

Importantly, there are no cracks at the interface between the initial and processed layers. This means good adhesion (at a cohesion level) of the modified surface layer to the base metal.

Tables 3.5 and 3.6 summarize the results of the X-ray microanalysis of the modified layer and adjacent structural layer similar to the initial one.

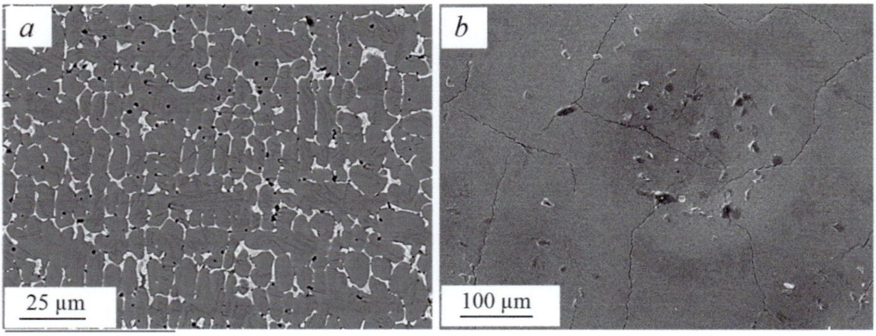

Fig. 3.22 SEM images of the plasma-surfaced 1.3355 layer: **a**—after 560°C annealing, **b**—after electron beam processing

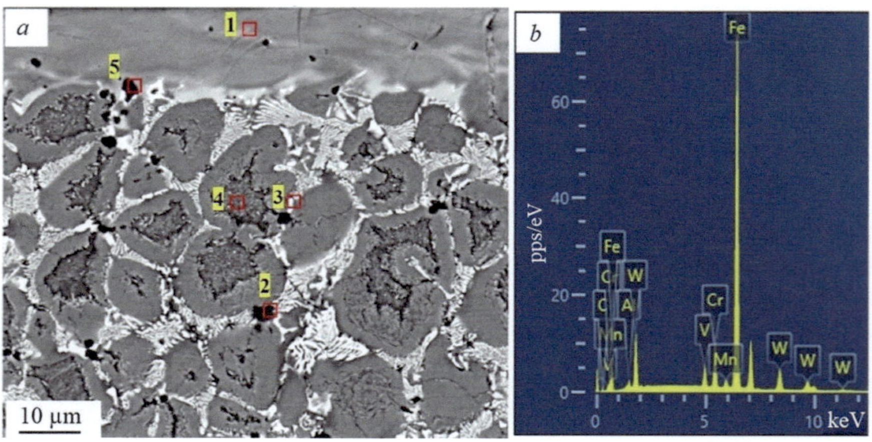

Fig. 3.23 Cross-sectional SEM image of the plasma-surfaced layer after tempering and electron beam processing (**a**) and EDS spectra obtained from the cross-sectional area (**b**). Highlighted regions indicate particular points analyzed. The arrow indicates the processed surface

According to these tables, the localized arrangement of *N* atoms is detected only in Al-rich inclusions (points 2 and 5) that suggests the presence of aluminum nitrides (oxynitrides). These inclusions (see Fig. 3.23a, dark color) are also observed in the layer modified by the electron beam. Second, the localized arrangement of C and Cr atoms is detected in the modified layer (point 1) and at its grain centers (point 4). Third, the increased content of W, Cr, and V atoms relative to the whole layer (see Table 3.6) is observed in interlayers separating grains in the plasma-surfaced layer (point 3).

EDS mapping visualizes places of preferred atomic arrangement in the hard-faced material. EDS maps in Fig. 3.24 show (1) the quasi-homogeneous distribution of chemical elements in the structure of the electron beam-processed layer, (2) Al-rich (probably oxynitrides) particles in this layer are 2 or 3 times smaller than those

Table 3.5 Elemental composition of the plasma-surfaced layer after high-temperature tempering and electron beam processing (see Fig. 3.23)

Elements	Point 1		Point 2		Point 3		Point 4		Point 5	
	wt.%	at.%	wt.%	at.%	wt.%	at.%	wt.%	at.%	wt.%	at.%
C	5.7	24.0	–	–	–	–	6.5	25.8	–	–
N	–	–	26.2	47.2	–	–	–	–	20.6	40.2
Al	1.0	1.9	44.9	41.9	7.3	16.1	1.2	2.2	45.2	45.8
Ti	–	–	–	–	0.6	0.8	–	–	–	–
V	3.0	3.0	1.9	0.9	7.1	8.3	1.9	1.8	2.2	1.2
Cr	4.8	4.7	2.7	1.3	4.0	4.7	3.9	3.5	2.2	1.2
Mn	0.5	0.4	–	–	–	–	–	–	–	–
Fe	67.4	61.1	17.2	7.7	58.5	62.8	74.1	63.5	21.3	10.4
W	17.5	4.8	7.1	1.0	22.5	7.3	12.4	3.2	8.5	1.3

Table 3.6 Elemental composition of the whole area of the electron beam-processed layer (see Fig. 3.23a)

Elements	Spectral lines	wt.%	σ, wt.%	at.%
C	K series	7.7	0.23	30.3
Al	K series	1.3	0.04	2.3
V	K series	2.6	0.02	2.4
Cr	K series	4.8	0.03	4.4
Mn	K series	0.5	0.2	0.4
Fe	K series	65.8	0.19	55.7
W	L series	17.3	0.11	4.5

in the underlying layers, (3) relatively lower C content in the modified layer, (4) the higher content of Cr, W, and V atoms in interlayers of the secondary phase, which locate along Fe-rich grain boundaries.

Thin foils were prepared from the specimen cross-section for STEM observations of the defect structure, elemental, and phase compositions. It was found that the modified layer consisted of columnar (cellular) rapidly solidified structure (Fig. 3.25) with the transverse cell size of 100–250 nm. The transverse size of secondary-phase particles locating along the boundaries of crystallization areas was 10–15 nm. There were also 45 nm faceted or globular particles on the boundaries and in the volume of columnar grains. Secondary-phase particles 5–10 nm in size also occurred in the volume of columnar grains.

TEM images obtained together with the microdiffraction patterns (Fig. 3.26) show that the columnar (cellular) structure is characterized by the small-angle (1.3–1.8 degrees) misorientation. This suggests that rapid solidification results in the formation of ferrite grains in the structure of the plasma-surfaced layer and the formation of

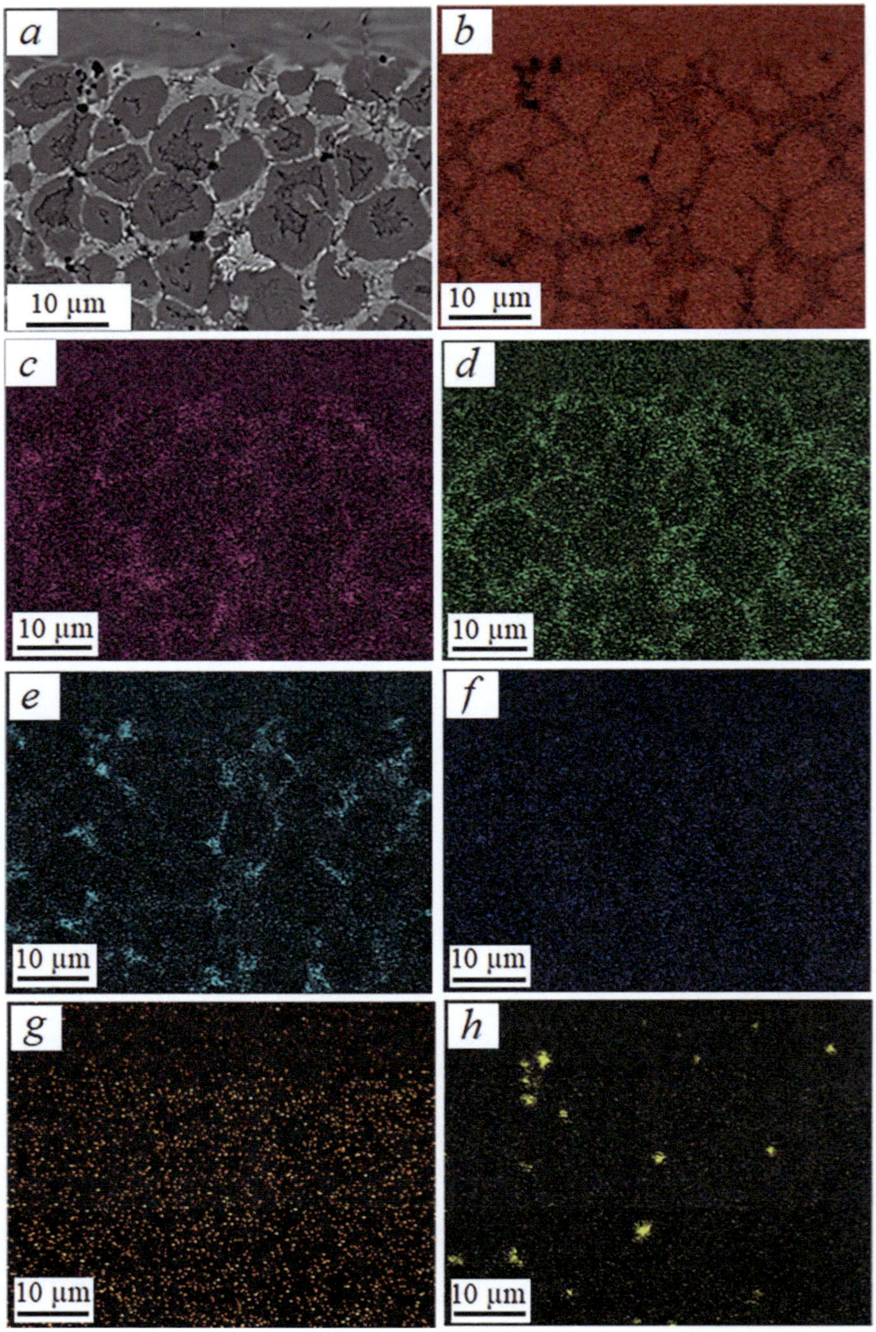

Fig. 3.24 SEM image of plasma-surfaced layer structure after electron beam processing: **a**—SEM image, **b**–**h**—EDS elemental mapping obtained in characteristic X-rays: **b**—Fe, **c**—Cr, **d**—W, **e**—V, **f**—Mn, **g**—C, **h**—Al atoms

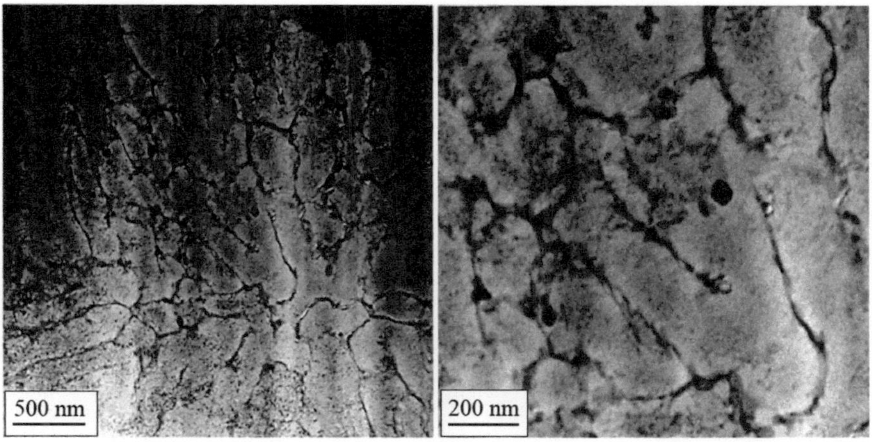

Fig. 3.25 STEM images of the plasma-surfaced layer after electron beam processing

nanosized particles of the secondary phase along the boundaries of columnar grains and inside them.

The elemental composition of the plasma-surfaced layer after electron beam processing was studied by the X-ray microanalysis. The results are presented in Table 3.7 and Fig. 3.27 and are in agreement with those given in Table 3.5 and indicate that secondary-phase inclusions along the boundaries of the columnar structure are rich in C, V, Cr, and W atoms.

The transmission electron microscopy was used to investigate the microstructure and phase composition of the modified layer (Fig. 3.28). It was found that the crystallization areas are produced by the BCC α-Fe-based solid solution. Nanosized particles of the secondary phase located in their volume are M_6C and $M_{23}C_6$ carbide particles (where M denotes Cr, Fe, and W elements expected in these carbides). In some cases, SAED patterns identified M_3C and Cr_7C_3 carbides.

Microhardness measurements were performed for the tempered 1.3355 layer deposited onto the ČSN 14331 steel after its additional tempering and electron beam processing. It was shown that the additional irradiation with the pulsed electron beam did not change the layer hardness, which was 5.3 GPa in both cases.

Tribology tests of the obtained layer after tempering and electron beam processing showed that the wear rate reduced by 3 times and higher (increased the wear resistance of the material) as well as the friction coefficient relative to tempered specimens not irradiated with the electron beam.

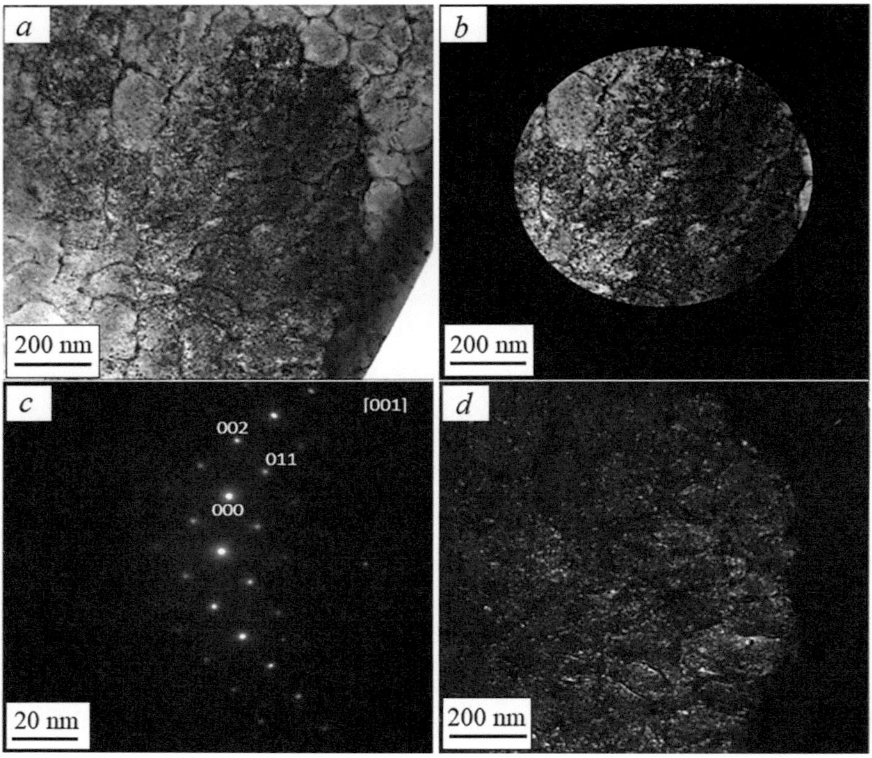

Fig. 3.26 TEM images of the plasma-surfaced layer after electron beam processing: **a**, **b**—bright-field images, **c**—SAED pattern obtained from the foil fragment (**b**), **d**—dark-field image in the [002] diffraction reflection of α-Fe

Table 3.7 Elemental composition of the foil fragment (Fig. 3.27) in particular points analyzed

Spectra	C	Al	Si	V	Cr	Mn	Fe	Mo	W
1	15.48	0.53	− 0.37	1.38	5.00	0.40	57.21	0.43	19.93
2	17.11	0.39	− 0.86	4.03	7.28	0.34	39.77	0.40	31.54
3	17.21	0.94	0.17	1.39	4.02	0.15	57.95	0.73	17.43
4	21.47	0.83	0.06	1.53	3.43	0.47	55.38	0.38	16.44
5	15.61	0.69	− 0.10	1.54	3.74	0.41	60.58	− 0.13	17.65
6	19.61	0.79	0.28	0.71	2.92	0.32	61.56	0.11	13.70
7	18.67	0.77	0.00	1.26	3.01	0.48	61.12	0.42	14.28
8	18.53	0.68	0.10	1.25	2.97	0.37	60.90	− 0.05	15.26
9	22.69	0.85	− 0.23	1.30	3.01	0.11	57.59	0.01	14.66
10	24.67	1.15	0.23	1.39	2.94	0.22	55.91	0.66	12.84

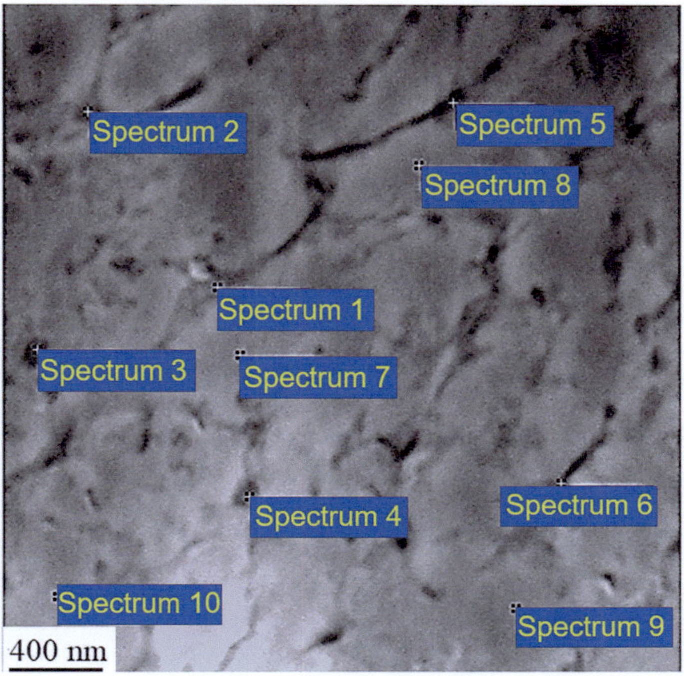

Fig. 3.27 STEM image of the foil fragment prepared from plasma-surfaced layer irradiated by electron beam

3.4 Conclusions

The elemental and phase compositions, defect structure, and mechanical and tribological properties of the plasma-surfaced 1.3355 layer were investigated when deposited onto a ČSN 14331 cast roll in the shielding nitrogen atmosphere, after high-temperature tempering and electron beam processing at a high-speed melting of the thin surface layer. The following conclusions have been drawn.

- Electron beam processing of the plasma-surfaced layer led to the formation of a thin (30–50 μm) polycrystalline surface layer. The grain volume of that layer consisted of the columnar (cellular) rapidly solidified structure with the cell size of 100–250 nm.
- The grain volume was formed by the BCC α-Fe-based solid solution. The transverse size of secondary-phase particles locating along the boundaries of crystallization areas was 10–15 nm. There were also 45 nm faceted or globular particles on the boundaries and in the volume of columnar grains. Secondary-phase particles 5–10 nm in size also occurred in the volume of columnar grains.

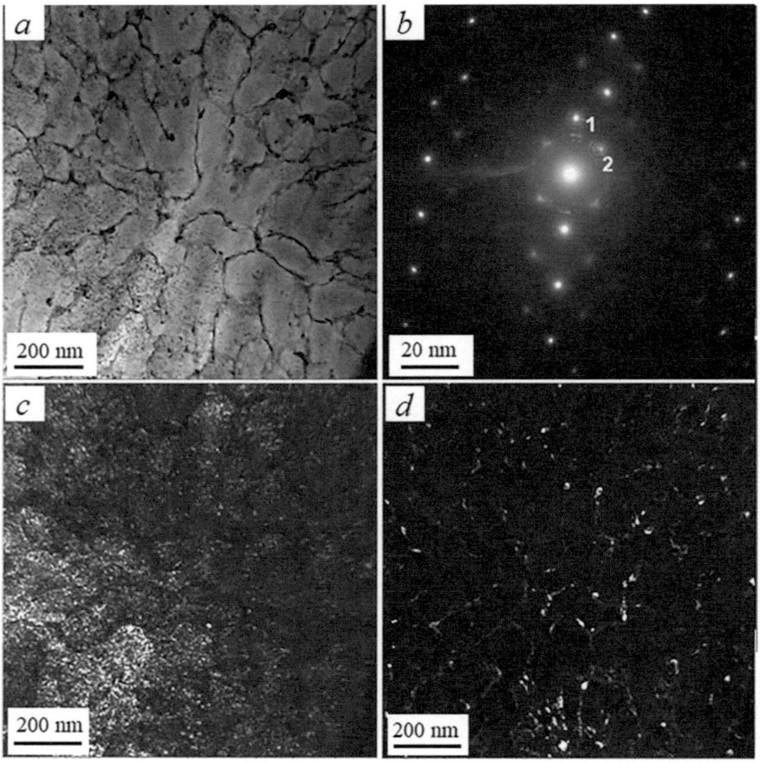

Fig. 3.28 TEM images of the plasma-surfaced layer after electron beam processing: **a**—bright-field images, **b**—SAED pattern (1 and 2—dark-field reflections for (**c**) and (**d**), respectively), **c**—dark-field image in [002] α-Fe + [331] M_6C diffraction reflections, **d**—dark-field image in [110] α-Fe + [422] $M_{23}C_6$ diffraction reflections

- Nanosized particles of the secondary phase located in the volume of crystallization areas were M_6C and $M_{23}C_6$ carbide particles (where M denotes Cr, Fe, W elements expected in these carbides). In some cases, SAED patterns identified M_3C and Cr_7C_3 carbides.
- Microhardness measurements performed for the tempered 1.3355 layer deposited onto the ČSN 14331 steel after its additional multiple tempering and electron beam processing showed that the wear rate reduced by 3 times and higher (the material wear resistance increased), while the microhardness remained unchanged (5.3 GPa).

References

1. N.N. Malushin, D.V. Valuev, V.L. Osetkovskii, S.A. Solodskii, in *Hard-Facing of Mining Machinery Parts with High Hardness Steels* (TPU, Tomsk, 2015)
2. N.N. Malushin, V.E. Gromov, D.A. Romanov, L.P. Bashchenko, in *Plasma-assisted Hardening of Heat-Resistant Alloys in Nitrogen* (Poligrafist, Novokuznetsk, 2022). (in Russian)
3. S.V. Raikov, V.E. Kormyshev, V.E. Gromov, Y.F. Ivanov, S.V. Konovalov, in *Wear-Resistant Plasma Surfacing: Structure, Phase Composition and Properties* (Novokuznetsk, 2017). (in Russian)
4. I.A. Ryabtsev, I.K. Senchenkov, *Theory and Practice of Steel Surfacing* (Ekotekhnologiya, Kyiv, 2013). (in Russian)
5. I.K. Pokhodnya, V.N. Shlepakov, S.Y. Maksimov, I.A.E.O. Ryabtsev, Paton electric welding institute research and development of arc welding and plasma-jet hard-facing with cored filler wire. Avtomaticheskaya svarka **12**(692), 34–42 (2010)
6. K.N. Vdovin, O.A. Nikitenko, N.A. Feoktistov, D.A. Gorlenko, Influence of nitrided ferrovanadium on microstructure parameters of cast steel/Hadfield steel products. Liteishchik Rossii **3**, 23–27 (2018)
7. S.P. Nefe'ev, A.N. Emelyushin, Plasma surfacing with cored filler of wear-resistant coatings. Vestnik Yugorskogo gosudarstvennogo universiteta **3**, 33–45 (2021)
8. S.N. Kutepov, A.A. Kalinin, A.E. Gvozdev, Modern steels for high-speed machining of metal alloys. Izvestiya Tul'skogo gosudarstvennogo universiteta. Tekhnicheskie nauki **10**, 597–607 (2018)
9. B.K. Rakhadilov, W. Wieleba, M.K. Kylyshkanov, A.B. Kenesbekov, M. Maulet, Structure and phase composition of high-speed steels. Bulletin of the Karaganda University. Physics **2**(98), 83–92 (2020)
10. V.P. Kireev, HSS hardening by dynamic microalloying and its effect on cutting tool wear resistance. Izvestiya Samarskogo nauchnogo tsentra Rossiiskoi akademii nauk **17**, 6(2), 414–418 (2015)
11. B.K. Rakhadilov, L.G. Zhurerova, M. Scheffler, A.K. Khassenov, Change in high-temperature wear resistance of high-speed steel by plasma nitriding. Bulletin of the Karaganda University. Physics **3**(91), 59–65 (2018)
12. Q. Ouyang, P. Luo, F. Zhang, Q. He, Y. Wang, S. Li, Analyzing the effect of CeB_6 on Microstructure and mechanical properties of high-speed steel consolidated by powder metallurgy. J. Mater. Eng. Perform. **27**, 5973–5983 (2018)
13. C.-C. Ding, M.-D. Zhao, Z.-D. Li, Y.-G. Cao, Effect of quenching temperature on microstructure and properties of high-speed axle steel. Cailiao Rechuli Xuebao/Trans. Mater. Heat. Treat. **39**(12), 49–56 (2018). https://doi.org/10.13289/j.issn.1009-6264.2018-0311
14. N.U. Rahman, L. Capuano, van der Meer A., M. de Rooij, D.T.A. Matthews, G. Walmag, M. Sinnaeve, A. García-Junceda, M. Castillo, G.R.B.E. Römer, Development and characterization of multilayer laser cladded high speed steels. Addit. Manuf. **24**, 76–85 (2018). https://doi.org/10.1016/j.addma.2018.09.009
15. G. Matlygin, A. Savilov, A. Nikolaev, S. Timofeev, Investigation of form deviations of high-speed steel (HSS) products under turning-milling operation using automatically programmed tools. Sci. Intensive Technol. Mech. Eng. 15–23 (2023). https://doi.org/10.30987/2223-4608-2023-7-15-23
16. L. Petrova, A. Sergeeva, V. Vdovin, Modification of a high-speed cutting tool surface by combined tungsten steel and nitrogen saturation. Sci. Intensive Technol. Mech. Eng. 24–32 (2023). https://doi.org/10.30987/2223-4608-2023-7-24-32
17. F.R. Egerton, *Physical Principles of Electron Microscopy* (Springer, Basel, 2016)
18. C.S.S.R. Kumar, in *Transmission Electron Microscopy Characterization of Nanomaterials* (Springer, New York, 2014)
19. C.B. Carter, D.B. Williams (eds.), *Transmission Electron Microscopy* (Springer, Berlin, 2016)
20. Yu.F. Ivanov, V.E. Gromov, S.V. Konovalov, V.E. Kormyshev et al., Structure and properties of strengthening layer on Hardox 450 steel. Mater. Sci. Technol. **33**(17), 2040–2045 (2017)

21. S.V. Konovalov, V.E. Kormyshev, V.E. Gromov, Yu.F. Ivanov, E.V. Kapralov, Phase composition and defect substructure of double surfacing, formed with V–Cr–Nb–W powder wire on steel. Inorg. Mater. Appl. Res. **8**(2), 313–317 (2017)
22. S.V. Konovalov, V.E. Kormyshev, V.E. Gromov, Yu.F. Ivanov, E.V. Kapralov, Gradient structure generated in Hardox 450 steel with built-up layer. Inorg. Mater. Appl. Res. **9**(3), 427–432 (2018)
23. V.G. Kurdyumov, L.M. Utevskii, R.I. Entin, *Transformations in Iron and Steel* (Nauka, Moscow, 1977). (in Russian)
24. E.J. Fasiska, H. Wagenblat, Dilatation of alpha-iron by carbon. Trans. Met. Soc. AIME **239**(11), 1818–1820 (1967)
25. N.N. Koval, Yu.F. Ivanov (eds.), *Evolution of Steel Surface Structure after Electron-Beam, Plasma-Ion Assisted Deposition* (Scientific Technology Publishing House, Tomsk, 2016). (in Russian)
26. V.E. Kormyshev, Yu.F. Ivanov, S.V. Konovalov, V.E. Gromov, A.D. Teresov, Electron beam modification of boronized layer on steel HARDOX-450 using cored filler wire. Fundamental'nye problemy sovremennogo materialoved. **13**(3), 382–387 (2016)
27. V.E. Kormyshev, V.E. Gromov, Yu.F. Ivanov, V.A. Bataev, O.A. Kondratova, E.N. Nikitina, Improvement of steel surface properties by cladding and subsequent electron beam treatment. Aktual'nye problemy v mashinostroenii **4**(1), 119–125 (2017)
28. V.E. Gromov, V.E. Kormyshev, A.M. Glezer, Yu.F. Ivanov, S.V. Konovalov, A.D. Teresov, Structure and tribological properties of the surface layer deposited onto martensitic steel and modified by electron beam processing. Fundamental'nye problemy sovremennogo materialoved. **14**(1), 124–129 (2017)
29. V.E. Gromov, Yu.F. Ivanov, A.M. Glezer, V.E. Kormyshev, S.V. Konovalov, Electron-beam modification of a surface layer deposited on low-carbon steel by means of arc spraying. Bull. Russ. Acad. Sci. Phys. **81**(11), 1353–1359 (2017)
30. V.E. Kormyshev, V.E. Gromov, Yu.F. Ivanov, S.V. Konovalov, A.D. Teresov, Structure and properties of hard-faced layer on low-carbon steel after electron beam processing. Izv. Rossiiskoi akademii nauk. Metally **4**, 46–51 (2017)

Chapter 4
Formation of Graded Structure and Phase Composition of Plasma-Surfaced HSS

4.1 Introduction

Modern engineering traditionally utilizes plasma surfacing of different wear-resistant materials [1]. Fe-based heat-resistant alloys with the high (~ 17–18%) tungsten and carbon (0.76–0.82%) content and possessing high hardness and wear resistance are the most attractive hard-facing material [2]. Directed-energy deposition of such alloy coatings is accompanied by cold cracking, which is eliminated by preliminary and additional high-temperature annealing and retarding cooling of parts [3]. This leads to the low coating hardness and wear resistance. In order to improve these parameters, the additional thermal treatment is required in accordance with the annealing–quenching–tempering procedure, which significantly restricts the application of these alloys [4]. In this regard, it is expedient to create plasma surfacing techniques for heat-resistant alloys, which permits to avoid cracking and retain high mechanical and tribological properties without additional heat treatments. This problem solution requires information about the formation of graded structures and phase compositions in plasma-surfaced materials. Solidification processes play a fundamental role for this formation [5] and thus determine the structure and mechanical properties of the hard acing.

In recent years, research on mechanisms and models of the material solidification on the surface of various geometry has become very popular [5–10]. Shown is the formation of either a cellular or dendritic structure or both of them, depending on external conditions (cooling rate, rotation speed, ambient temperature, and undercooling) [5, 6]. According to [7, 8], the main formation mechanism of these structures is the instability of crystallization front caused by the temperature fall of the phase transition due to the melt substitution for the impurity and solutal undercooling. Thus, the shape of the phase boundary determines the impurity distribution in a crystal [9]. The main conclusion of all theories of morphological stability is a loss of stability of the crystallization front relative to small perturbations at a certain ratio between the temperature and solute gradient [10]. This leads to the formation of

A. I. Potekaev et al., *High-Speed Steel After Tempering and Electron Beam Processing*, SpringerBriefs in Applied Sciences and Technology,
https://doi.org/10.1007/978-981-95-1398-7_4

complex structures and phase compositions and promotes the particle nucleation in dissolved impurities. Thus, an extended area of the phase transition appears prior to the crystallization front [11]. Models proposed in [5–11] consider only the motion of phase transition boundaries and do not allow for the motion of heat penetration boundaries of the melt. It is usually assumed that stabilized temperature distribution occurs at this boundary or it is directed into infinity [12, 13]. Consideration of the fact that the heat penetration boundary is not infinitely large according to the thermal solution [14], leads to a more rapid particle growth as compared to the problem solution regardless of this fact. Thus, in constructing mathematical models of the plasma effect on the material structure, it is necessary to consider the boundary motion of both phase transition and heat penetration.

A search for the formation mechanisms of the graded structure and phase composition of heat-resistant alloys during their plasma surfacing onto the rotating cast roll must be based on the understanding and development of the Mullins–Sekerka instability. Its analysis allows to determine conditions in which this structure and phase composition arise with account for the motion of the heat penetration boundary. Works of [15–20] focus on studying the Mullins–Sekerka instability. In [15, 16], it is studied for binary alloys. They formulate criteria for the absolute and relative stability of a spherical crystallization nucleus for these alloys and show that with the particle growth, the initial concentration in a molten binary alloy at first suppresses and then increases the particle morphological stability. They also determine the critical concentration for the initiation of this effect. Gildas and Gandin [17] study a spherical particle with respect to nonstationary members of diffusion equations and full record of the diffusion coefficient matrix that show that the stability criteria are not reduced to a low speed, as it shown in [18]. The stability of the growing sphere can be considered at a high growth rate of perturbations, which also matches the high oversaturation. Research findings clearly show that threshold values for the interface destabilization of the growing sphere largely depend on the growth rate. It is found that the degree of spherical harmonic at which stability is not supported grows with increasing growth rate. Works [16–18] pay insufficient attention to the surface tension of the melt, while it can substantially shift the wavelength, which the maximum perturbation mode falls on [19, 20]. Gus'kov [19] detects the wavelength dependence on the surface tension and mechanism of the crystal growth. It is shown that most of binary compositions considered by Gus'kov satisfy the crystal growth via screw dislocations, and the dependence $\lambda - V^{-1/2}$ is linear and coincides with the experimental data. According to Gus'kov and Orlov [20], the surface tension significantly affects the kinetic coefficient via screw dislocations. Therefore, in mathematical simulations of the material solidification during plasma processing and in formulation of stability criteria for the crystal–melt interface, it is necessary to account for the surface tension and crystal growth in addition to solutal undercooling. The study of the Mullins–Sekerka instability includes the following stages: (1) determination of the perturbation behavior at the crystal–melt interface and its curvature effect on the liquidus temperature, (2) calculation of temperature fields and concentration of solid and liquid phases, and (3) calculation of the growth rate of perturbations based on the boundary conditions of the phase transition.

4.2 Stefan Problem Solution for Two Moving Boundaries

Let us consider the directed crystallization front along the solid axis r. In contrast to traditional works [21–23] concerning this problem, which suggest that the crystal growth is proportional to $t^{1/2}$, we do not use this hypothesis. The crystal growth in time is traced by the solution of a system of kinetic equations obtained from the balance conditions of temperature and substance on the boundary of phase transitions. The geometry of the problem is illustrated in Fig. 4.1. The initial phase occupies the range $R_2(t) < r < +\infty$ and has the temperature T_0. The second phase appears in approaching to temperature T_{**} and occupies the range $R_1(t) < r < R_2(t)$. The third phase starts at T_* temperature and occupies the range $0 < r < R_1(t)$. The heat-transfer equation for each region is written as

$$\begin{aligned} \frac{\partial T_1}{\partial t} &= \chi_1\left(\frac{1}{r}\frac{\partial}{\partial r}\left(r\frac{\partial T_1}{\partial r}\right)\right), \quad R_2(t) < r < +\infty, \\ \frac{\partial T_2}{\partial t} &= \chi_2\left(\frac{1}{r}\frac{\partial}{\partial r}\left(r\frac{\partial T_2}{\partial r}\right)\right), \quad R_1(t) < r < R_2(t), \\ \frac{\partial T_3}{\partial t} &= \chi_3\left(\frac{1}{r}\frac{\partial}{\partial r}\left(r\frac{\partial T_3}{\partial r}\right)\right), \quad 0 < r < R_1(t), \end{aligned} \tag{4.1}$$

where χ_1, χ_2, χ_3 are thermal conductivities in regions 1–3. The moving boundaries of phase transitions are determined by the balance equation of temperature and thermal flows:

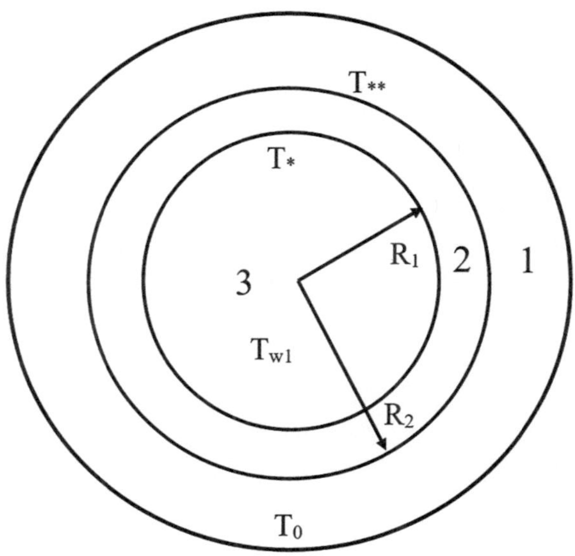

Fig. 4.1 Geometry of the problem

$$
\begin{aligned}
&T_3 = T_2 = T_*, r = R_1(t),\\
&\left(-\lambda_3\frac{\partial T_3}{\partial r} + \lambda_2\frac{\partial T_2}{\partial r}\right) = \Delta H_1\frac{\mathrm{d}R_1}{\mathrm{d}t}, r = R_1(t),\\
&T_2 = T_1 = T_{**}, r = R_2(t),\\
&\left(\lambda_1\frac{\partial T_1}{\partial r} - \lambda_2\frac{\partial T_2}{\partial r}\right) = \Delta H_2\frac{\mathrm{d}R_2}{\mathrm{d}t}, r = R_2(t),
\end{aligned}
\tag{4.2}
$$

where λ_1, λ_2, λ_3 are thermal conductivities in in regions 1–3, and ΔH_1 and ΔH_2 are heat values of phase transitions. At $r \to 0$, the temperature is T_{w1}, and at $r \to \infty$, it is T_0. The initial conditions take the form

$$
T_2(r, 0) = T_0, R_1(0) = R_0,\ R_2(0) = R_0^*, \tag{4.3}
$$

where R_0 and R_0^* are initial radii of phase transition boundaries. The system of (4.1)–(4.3) equations is solved as

$$
T_i(r, t) = A_i + B_i E_i\left(\frac{r^2}{4\chi_i t}\right), \tag{4.4}
$$

where A_i, B_i are arbitrary constants, and $E_i(z)$ is the integral exponent, $i = 1$–3. Substituting (4.4) in the boundary conditions (4.2) and initial conditions (4.3), we get

$$
\begin{aligned}
&T_1(r, t) = T_0 - (T_0 - T_{**})\frac{E_i\left(\frac{r^2}{4\chi_1 t}\right)}{E_i\left(\frac{R_2^2}{4\chi_1 t}\right)}, \quad R_2(t) < r < +\infty,\\
&T_2(r, t) = \frac{T_{**}E_i\left(\frac{R_1^2}{4\chi_2 t}\right) - T_* E_i\left(\frac{R_2^2}{4\chi_2 t}\right) + (T_* - T_{**})E_i\left(\frac{r^2}{4\chi_2 t}\right)}{E_i\left(\frac{R_1}{4\chi_2 t}\right) - E_i\left(\frac{R_2}{4\chi_2 t}\right)},\\
&\qquad R_1(t) < r < R_2(t),\\
&T_3(r, t) = \frac{T_* E_i\left(\frac{R^2}{4\chi_3 t}\right) - T_{w1}E_i\left(\frac{R_1^2}{4\chi_3 t}\right) - (T_* - T_{w1})E_i\left(\frac{r^2}{4\chi_3 t}\right)}{E_i\left(\frac{R^2}{4\chi_3 t}\right) - E_i\left(\frac{R_1^2}{4\chi_3 t}\right)},\\
&\qquad 0 < r < R_1(t).
\end{aligned}
\tag{4.5}
$$

The parameter R, having the measure of the radial coordinate, is selected to eliminate divergence at $r \to 0$. It is supposed to be 10^{-8} m. Substituting (4.5) in the heat balance equation at the boundary of phase transitions, we get the following equations:

$$\frac{\mathrm{d}R_1}{\mathrm{d}t} = \frac{F_1 \exp\left(-\frac{R_1^2}{4\chi_1 t}\right)}{R_1\left(E_i\left(\frac{R_1^2}{4\chi_2 t}\right) - E_i\left(\frac{R_2^2}{4\chi_2 t}\right)\right)} - \frac{F_2 \exp\left(-\frac{R_1^2}{4\chi_3 t}\right)}{\left(E_i\left(\frac{R_1^2}{4\chi_2 t}\right) - E_i\left(\frac{R_2^2}{4\chi_2 t}\right)\right)R_1},$$

$$\frac{\mathrm{d}R_2}{\mathrm{d}t} = \frac{G_1 \exp\left(-\frac{R_2^2}{4\chi_2 t}\right)}{R_2\left(E_i\left(\frac{R_1^2}{4\chi_2 t}\right) - E_i\left(\frac{R_2^2}{4\chi_2 t}\right)\right)} - \frac{G_2 \exp\left(-\frac{R_2^2}{4\chi_1 t}\right)}{\left(E_i\left(\frac{R_2^2}{4\chi_1 t}\right)\right)R_2},$$

$$F_1 = \frac{2(T_{**} - T_*)\lambda_2}{\Delta H_1}, F_2 = \frac{2(T_* - T_{w1})\lambda_3}{\Delta H_1}, G_1 = \frac{2(T_* - T_{**})\lambda_2}{\Delta H_2},$$

$$G_2 = \frac{2(T_0 - T_{**})\lambda_1}{\Delta H_2}. \tag{4.6}$$

(4.6) is the system of differential equations, which is solved by the high-order Runge–Kutta methods. For convenience, let us use nondimensional variables $\tilde{R}_i = R/R_0$ and $\tau = (\chi_1/R_0^2)t$. Since at $t \to 0$, $E_i\left(R_i^2/4\chi t\right) \to 0$, the time is set to be 10^{-9} s. Table 4.1 presents parameters of the iron–tungsten system.

In region 1, the melt temperature is $T_0 = 1811$ K. On the boundary R_2, the liquidus temperature $T_{**} = 1806$ K, while on the boundary R_1, the solidus temperature $T_* = 1803$ K. These values are derived from the phase diagram [19] for 18 wt.% W. At $T_{w1} < T_*$, the temperature is 1790 K. It is assumed that $\chi_2 = \chi_3$ and $\lambda_2 = \lambda_3$, $\Delta H_1 = \Delta H_2$. Figure 4.2 shows the kinetics of the interface motion. At $R_0 = 1$ μm, the boundary coordinate R_1 grows linearly up to $\tau = 0.028$ (4.4118 ns), while the boundary coordinate R_2 changes nonmonotonically and drops after $\tau > 0.028$ (4.4118 ns). If $R_0 = 10$ μm, then the same tendency occurs, but the solidification process takes 41.176 μs.

The obtained results show that with decreasing nuclei size, the time of their stable growth decreases by 4 times. A drastic decrease in the radial coordinate R_2 and a drastic growth in the radial coordinate R_1 indicate to the crystallization front instability caused by the surface tension and solutal undercooling. Rapid solidification in small regions is explained by the presence of the high surface energy, which tends to decrease via the size growth and changes in the interface boundary configuration [24].

Table 4.1 Parameters of the iron–tungsten system

Melt temperature T_{w1}	1811 K
Heat of melting ΔH	270 kJ/kg
Density ρ_1	6980 kg/m^3
Thermal conductivity λ_1 in region 1	39 W/m · K
Thermal conductivity λ_2 and λ_3 in regions 2 and 3	35 W/m · K
Thermal conductivity χ_1 in region 1	6.8×10^{-6} m^2/s
Thermal conductivity χ_2 in regions 2 and 3	6.9×10^{-6} m^2/s

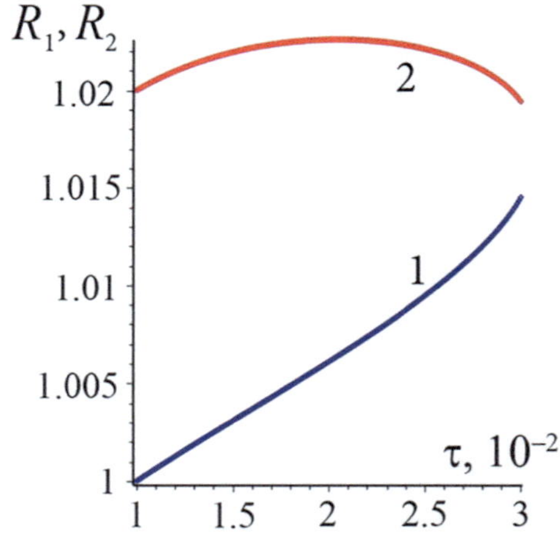

Fig. 4.2 Time dependences of radial coordinates of the crystallization front at T_{w1} = 1790 K, T_* = 1803 K, T_{**} = 1806 K, T_0 = 1811 K, R_0 = 1 μm: 1—interface between phases 1 and 2, 2—interface between phases 2 and 3

In Fig. 4.3, time dependences are presented for the nondimensional radial coordinate. One can see that the size of region 2 reduces (curves 2 and 3), and the size of region 3 grows with increasing time.

Another tendency is observed in Fig. 4.4 at T_0 = 1790 K, T_{**} = 1803 K, T_* = 1803 K, and T_{w1} = 1811 K. According to Fig. 4.4a, the interface coordinates R_1 and R_2 reduce linearly and parabolically, respectively. According to Fig. 4.4b, the size of the regions 2 and 3 (curves 2 and 3) reduces.

Theoretical studies of the iron–tungsten system solidification based on kinetic Eqs. (4.6) show that the boundary R_2 of the liquidus temperature moves by a descending parabola, rather than by the dependence $R \sim t^{1/2}$, while the boundary R_1 of the solidus temperature moves linearly. Upon the certain time point, these boundaries approach to each other that indicates to stopping of the crystallization process or the development of its instability. The obtained temperature dependences

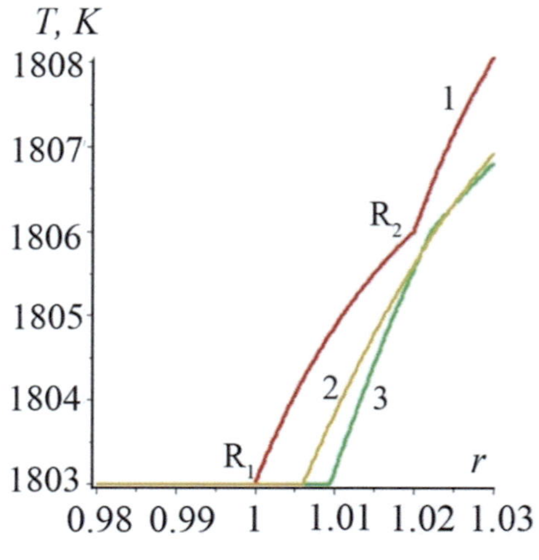

Fig. 4.3 Time dependences of radial coordinates at T_{w1} = 1790 K, T_* = 1803 K, T_{**} = 1806 K, T_0 = 1811 K, R_0 = 1 μm at different time: 1—$\tau = 0.01$, 2—$\tau = 0.02$, 3—$\tau = 0.025$

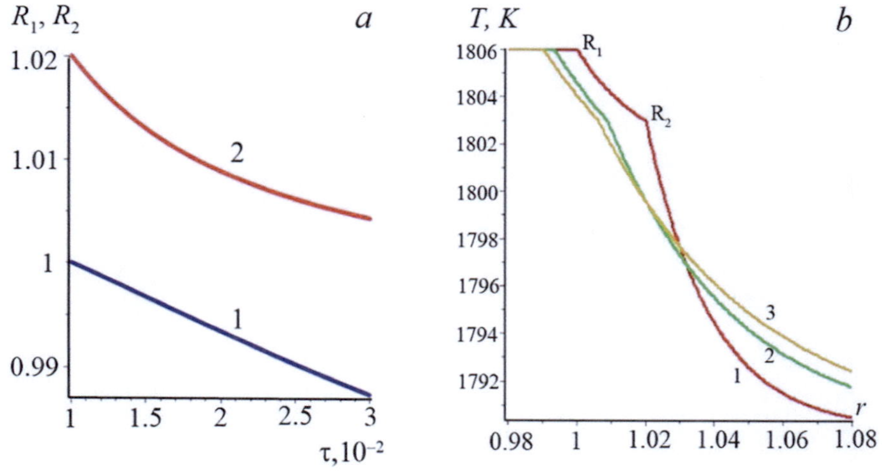

Fig. 4.4 Time dependences of radial coordinates at $T_{w1} = 1811$ K, $T_* = 1806$ K, $T_{**} = 1803$ K, $T_0 = 1790$ K, $R_0 = 1$ μm: **a**—1—interface between 1 and 2 phases, 2—interface between 2 and 3 phases, **b**—1—$\tau = 0.01$, 2—$\tau = 0.02$, 3—$\tau = 0.025$

can be used to study this instability. A further model development toward its adaptation to the plasma surfacing process considers the rotation of one of the media and more strict account for the content of alloying elements.

4.3 The Mullins–Sekerka Instability in the Formation of Crystal Structures

As mentioned above, the Mullins–Sekerka instability should be studied in several stages: (1) determination of the perturbation behavior at the interface and its curvature effect on the liquidus temperature, (2) calculation of temperature fields and concentration of solid and liquid phases, (3) calculation of the growth rate of perturbations based on the boundary conditions of the phase transition. This paragraph focuses on the first and second stages, assuming that the crystallization front is cylindrical. SEM observations are used to verify the obtained results of the plasma-surfaced layer structure.

Figure 4.5 presents the SEM image of the plasma-surfaced layer structure, which consists of the cellular and dendritic morphology. As shown in Fig. 4.5b, its grain size varies between 3 and 45 μm, while the most probable size is 10–15 μm. As can be seen from Fig. 4.5b, the instability of the crystallization front has one peak matching the wavelength of the most probable grain size. The presence of two morphologies allows assuming that there are two types of instability: soft (cellular structure) and hard (dendritic structure).

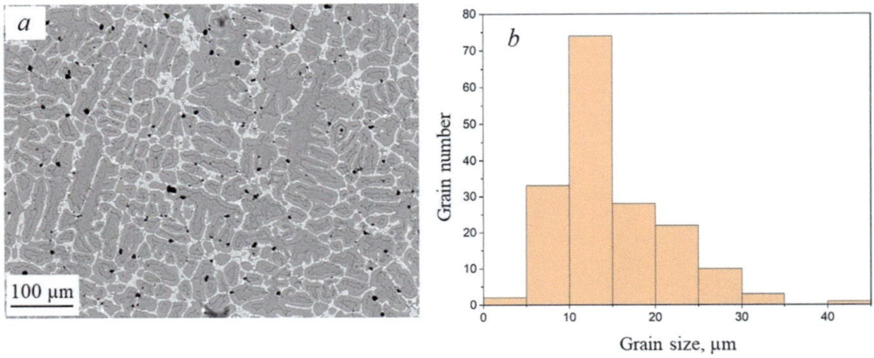

Fig. 4.5 SEM image of the plasma-surfaced layer structure (**a**) and grain size distribution (**b**)

Similar to the above paragraph, let us consider the stability of the cylindrical crystallization front relative to small harmonic disturbances (Fig. 4.6). For convenience, let us introduce the following nondimensional variables, as in [19]: $T_l = T_{lr}/T_0$, $T_s = T_{sr}/T_0$, $C_l = C_{lr}/C_0$, $r = r_r/a$, $z = z_r/a$, $t = \frac{\chi_0}{a^2}t_r$, $D_l = D_{lr}/\chi_0$, $\chi_l = \chi_{lr}/\chi_0$, $\chi_s = \chi_{sr}/\chi_0$, where T_{lr}, T_{sr} are temperatures of liquid and solid phases, C_{lr} is the impurity content in the liquid phase, r_r, z_r are radial and axial coordinates, D_{lr} is the impurity diffusion coefficient in a liquid, χ_{lr}, χ_{sr} are thermal conductivities in liquid and solid phases, respectively, T_0 is the temperature of phase transition (equal the liquidus temperature), C_0 is the initial impurity content, a is the initial radius (~ 1 μm) of cylindrical nucleus, and χ_0 is the scaling factor (~ 10^{-5} m^2/s) of thermal conductivity.

Latent heat ΔH of the phase transition is represented by $\varepsilon = \Delta H / c_0 T_0$, where c_0 is the heat capacity of the substance at the phase transition temperature. Equations of thermal conductivity and diffusion are written in nondimensional variables for solid and liquid phases as follows:

$$\begin{aligned}
\frac{\partial T_l}{\partial t} - V\frac{\partial T_l}{\partial r} &= \chi_l\left(\frac{\partial^2 T_l}{\partial r^2} + \frac{1}{r}\frac{\partial T_l}{\partial r} + \frac{\partial^2 T_l}{\partial z^2}\right),\\
\frac{\partial T_s}{\partial t} - V\frac{\partial T_s}{\partial r} &= \chi_s\left(\frac{\partial^2 T_s}{\partial r^2} + \frac{1}{r}\frac{\partial T_s}{\partial r} + \frac{\partial^2 T_s}{\partial z^2}\right),\\
\frac{\partial C_l}{\partial t} - V\frac{\partial C_l}{\partial r} &= D_l\left(\frac{\partial^2 C_l}{\partial r^2} + \frac{1}{r}\frac{\partial C_l}{\partial r} + \frac{\partial^2 C_l}{\partial z^2}\right).
\end{aligned} \quad (4.7)$$

The boundary conditions are

$$\begin{aligned}
&\chi_s\frac{\partial T_s}{\partial r} - \chi_l\frac{\partial T_l}{\partial r} = \varepsilon V,\, D_l\frac{\partial C_l}{\partial r} = (1 - k_s)C_l V,\, r = a + \xi(r, z),\\
&T_l = T_s,\, V = V(T, C),\, r = a + \xi(r, z),\\
&C_l = 1,\, T_l = T_{00},\, r \to \infty.
\end{aligned} \quad (4.8)$$

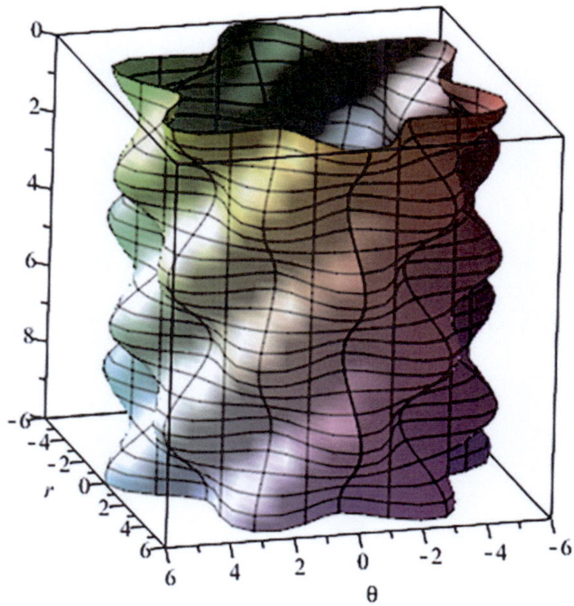

Fig. 4.6 Problem geometry of instability of cylindrical crystallization front

Let us write (4.7) as a sum of stationary and disturbed component to study the crystallization front stability:

$$T_l = T_{0l}(r) + T_{l1}(r,z,t),\ \ T_s = T_{0s}(r) + T_{s1}(r,z,t),\ \ C_l = C_{0l}(r) + C_{l1}(r,z,t),$$
$$V = V_s + V_1(r,z,t). \tag{4.9}$$

For the stationary component:

$$\begin{aligned}&\frac{\partial^2 T_{0l}}{\partial r^2} + \left(\frac{1}{r} + \frac{V_s}{\chi_1}\right)\frac{\partial T_{0l}}{\partial r} = 0,\ \frac{\partial^2 T_{0s}}{\partial r^2} + \left(\frac{1}{r} + \frac{V_s}{\chi_2}\right)\frac{\partial T_{0s}}{\partial r} = 0,\\ &\frac{\partial^2 C_{0l}}{\partial r^2} + \left(\frac{1}{r} + \frac{V_s}{D}\right)\frac{\partial C_{0l}}{\partial r} = 0.\end{aligned} \tag{4.10}$$

With regard to (4.9), the boundary conditions for (4.10) take the form

$$
\begin{aligned}
&T_{0l} = T_{0s},\ r = a,\\
&\chi_s \frac{\partial T_{0s}}{\partial r} - \chi_l \frac{\partial T_{0l}}{\partial r} = \varepsilon V_s,\ r = a,\\
&D_l \frac{\partial C_{0l}}{\partial r} = (1 - k_s) C_{0l} V_s,\ r = a,\\
&\chi_s \frac{\partial T_{0s}}{\partial r} = I_0, r = a,\\
&C_{0l} = 1, T_l = T_{00}, r \to \infty, T_{0s} = T_0, r = r_0.
\end{aligned}
\tag{4.11}
$$

Equations of the interface disturbance are written as

$$
\begin{aligned}
&\frac{\partial T_{l1}}{\partial t} - V_s \frac{\partial T_l}{\partial r} - V_1 G_l = \chi_l \left(\frac{\partial^2 T_{l1}}{\partial r^2} + \frac{1}{r} \frac{\partial T_{l1}}{\partial r} + \frac{\partial^2 T_{l1}}{\partial z^2} \right),\\
&\frac{\partial T_{s1}}{\partial t} - V_s \frac{\partial T_{s1}}{\partial r} - V_1 G_s = \chi_s \left(\frac{\partial^2 T_{s1}}{\partial r^2} + \frac{1}{r} \frac{\partial T_{s1}}{\partial r} + \frac{\partial^2 T_{s1}}{\partial z^2} \right),\\
&\frac{\partial C_{l1}}{\partial t} - V_s \frac{\partial C_{l1}}{\partial r} - V_1 G_c = D_l \left(\frac{\partial^2 C_{l1}}{\partial r^2} + \frac{1}{r} \frac{\partial C_{l1}}{\partial r} + \frac{\partial^2 C_{l1}}{\partial z^2} \right).
\end{aligned}
\tag{4.12}
$$

Therefore, the boundary conditions (4.8) take the form

$$
\begin{aligned}
&T_{l1} = T_{s1}, r = a,\\
&\chi_s \frac{\partial T_{s1}}{\partial r} - \chi_l \frac{\partial T_{l1}}{\partial r} = \varepsilon V_1, r = a,\\
&D_l \frac{\partial C_{l1}}{\partial r} = (1 - k_s)(V_s C_{l1} + C_{0l} V_1), r = a,\\
&C_{l1} = 0, T_{l1} = 0, r \to \infty, T_{s1} = 0, r \to 0.
\end{aligned}
\tag{4.13}
$$

The linear analysis of stability requires knowledge of the analytical form of undisturbed temperature gradients included in (4.12). To obtain them, the boundary problem (4.10), (4.11) must be solved:

$$
\begin{aligned}
&T_{0l} = T_{00} + \left(\frac{T_0 - T_{00}}{E_i\left(\frac{V_s a}{\chi_l}\right)} + \frac{\frac{I_0 a}{\chi_s} \exp\left(\frac{V_s}{\chi_s} a\right)\left(E_i\left(\frac{V_s}{\chi_s} r_0\right) - E_i\left(\frac{V_s}{\chi_s} a\right)\right)}{E_i\left(\frac{V_s a}{\chi_l}\right)} \right) E_i\left(\frac{V_s}{\chi_l} r\right),\\
&T_{0s} = T_0 + \frac{I_0 a}{\chi_s} \exp\left(\frac{V_s}{\chi_s} a\right)\left(E_i\left(\frac{V_s}{\chi_s} r_0\right) - E_i\left(\frac{V_s}{\chi_s} r\right)\right),\\
&C_{0l} = 1 + \frac{V_s a (1 - k_s) E_i\left(\frac{V_s}{D_l} r\right)}{V_s a (k_s - 1) E_i\left(\frac{V_s}{D_l} a\right) + D_l \exp\left(-\frac{V_s}{D_l} a\right)}.
\end{aligned}
\tag{4.14}
$$

Next, let us solve the boundary problems (4.12) and (4.13) for the disturbed temperature, concentration, and movement of the crystallization front. Its solution is the following:

$$\begin{aligned} &T_{l1}(r,z,t) = T_{m1}(r)\exp(\omega t + kz),\ T_{s1}(r,z,t) = T_{m2}(r)\exp(\omega t + kz),\\ &C_{l1}(r,z,t) = C_{m1}(r)\exp(\omega t + kz),\ V_1 = V_m(r)\exp(\omega t + kz), \end{aligned} \tag{4.15}$$

where $\omega = \omega_1 + i\omega_2$, $k = k_1 + ik_2$. Then (4.13) takes the form

$$\begin{aligned} &\frac{\mathrm{d}^2 T_{m1}}{\mathrm{d}r^2} + \left(\frac{1}{r} + \frac{V_s}{\chi_l}\right)\frac{\mathrm{d}T_{m1}}{\mathrm{d}r} + \frac{\left(\chi_l k^2 - \omega\right)}{\chi_l} T_{m1} = -\frac{V_m}{\chi_l} G_l,\\ &\frac{\mathrm{d}^2 T_{m2}}{\mathrm{d}r^2} + \left(\frac{1}{r} + \frac{V_s}{\chi_l}\right)\frac{\mathrm{d}T_{m2}}{\mathrm{d}r} + \frac{\left(\chi_s k^2 - \omega\right)}{\chi_s} T_{m2} = -\frac{V_m}{\chi_s} G_s,\\ &\frac{\mathrm{d}^2 C_m}{\mathrm{d}r^2} + \left(\frac{1}{r} + \frac{V_s}{D_l}\right)\frac{\mathrm{d}C_m}{\mathrm{d}r} + \frac{\left(D_l k^2 - \omega\right)}{D_l} C_m = -\frac{V_m}{D_l} G_c. \end{aligned} \tag{4.16}$$

Therefore, the boundary conditions (4.13) are written as

$$\begin{aligned} &T_{m1}(a) = T_{m2}(a),\\ &\chi_s T_{m2}(a)' - \chi_l T_{m1}(a)' = \varepsilon V_m(a),\\ &D_l C_m(a)' = (k_s - 1)(V_s C_m(a) + C_{0l} V_m(a)),\\ &C_m = 0,\ T_{m1} = 0,\ r \to \infty,\ T_{m2} = 0,\ r \to 0. \end{aligned} \tag{4.17}$$

Here, the prime symbol denotes the derivative with respect to the radial coordinate. Similar to [19], the dependence of the interface boundary perturbations on the maximum perturbation temperature and concentration is as follows:

$$V_m(a) = \theta T_{m1}(a) + \gamma C_{\mathrm{m}}(a), \tag{4.18}$$

where $\theta = \frac{\partial V}{\partial T}$ and $\gamma = \frac{\partial V}{\partial C}$. The system of Eq. (4.16) includes degenerate nonhomogeneous hypergeometric equations solved by Kummer's function. Derivation and analysis of the dispersion equation, which includes Kummer's function, is a complex and nontrivial problem. In regard to the latter point, reference will therefore be made only to special cases. In the first case, the convective member in (4.7) is neglected. Therefore, stationary solutions take the form

$$\begin{aligned} &T_{0l} = T_{00} + \frac{a\ln(r/b_1)(I_0 - \varepsilon V_s)}{\chi_l},\ T_{0s} = T_0 + \frac{a\ln(r/b)I_0}{\chi_s},\\ &C_{0l} = 1 + \frac{(k_s - 1)V_s a\ln(r/b_1)}{(1 - k_s)aV_s\ln(a/b_1) + D_l}. \end{aligned} \tag{4.19}$$

Consequently, (4.16) are written as follows:

$$\frac{d^2T_{m1}}{dr^2}+\frac{1}{r}\frac{dT_{m1}}{dr}-S_1^2T_{m1}=0,\ \frac{d^2T_{m2}}{dr^2}+\frac{1}{r}\frac{dT_{m2}}{dr}-S_2^2T_{m2}=0,$$
$$\frac{d^2C_m}{dr^2}+\frac{1}{r}\frac{dC_m}{dr}-S_3^2C_m=0. \tag{4.20}$$

where $S_1^2=\frac{\omega}{\chi_l}-k^2$, $S_2^2=\frac{\omega}{\chi_s}-k^2$, $S_3^2=\frac{\omega}{D_l}-k^2$.

The system of Eq. (4.20) is calculated as

$$T_{m1}=A_1I_0(S_1r)+A_2K_0(S_1r),\ T_{m2}=A_3I_0(S_2r)+A_4K_0(S_2r),$$
$$C_m=A_5I_0(S_3r)+A_6K_0(S_3r). \tag{4.21}$$

Substitution of (4.21) into the boundary conditions (4.17) and transformation with respect to (4.19) yields the dispersion equation:

$$\left(\frac{K_0(S_1a)I_1(S_2a)S_2}{I_0(S_2a)}\chi_2+\chi_1K_1(S_1a)S_1-\varepsilon\theta K_0(S_1a)\right)$$
$$\times((1-k_s)K_0(S_3a)(C_{0l}\gamma+V_s)-D_lK_1(S_3a)S_3)$$
$$+(1-k_s)\varepsilon\gamma C_{0l}K_0(S_3a)K_0(S_1a)\theta=0. \tag{4.22}$$

At the impurity content approximately equaling the eutectic content, ($1–k_s$) value tends to zero and thus can be neglected in (4.22). The resulting relation is

$$\frac{I_1(S_2a)}{I_0(S_2a)}S_2\chi_2+\frac{K_1(S_1a)}{K_0(S_1a)}S_1\chi_1-\varepsilon\theta=0. \tag{4.23}$$

Similar to [19], $\theta=\frac{\Lambda\omega}{\Lambda\Gamma k^2-\omega}$, where $\Lambda=\frac{\partial V}{\partial\Delta T}$ is the coefficient, depending on the mechanism of the crystal growth, $\Gamma=\alpha\Gamma_r$, where Γ_r is the ratio between the product of the surface tension by the phase transition temperature and phase transition latent energy, $\alpha=1/aT_0$. At $S_{1,2}>>1$, the approximated Bessel functions can be written as

$$I_0(S_2a)\approx I_1(S_2a)\approx \exp(S_2a)/\sqrt{2\pi S_2a}K_0(S_1a)\approx K_1(S_1a)\approx\pi\exp(-S_1a)/\sqrt{2\pi S_1a}.$$

Therefore, (4.23) takes the form

$$S_2\chi_2+S_1\chi_1-\varepsilon\theta=0. \tag{4.24}$$

Let us assume that $k_1=0$ and $\omega_2=0$, then $S_1=\sqrt{k_2^2+\frac{\omega_1}{\chi_l}}$, $S_2=\sqrt{k_2^2+\frac{\omega_1}{\chi_s}}$. Let us write $\omega_1=\frac{\delta V_s^2}{D_l}$, $k_2^2=\frac{V_s^2}{D_l^2}Y$, then $S_1=\frac{V_s}{D_l}\sqrt{Y+\frac{\delta D_l}{\chi_l}}$, $S_2=\frac{V_s}{D_l}\sqrt{Y+\frac{\delta D_l}{\chi_s}}$. If $Y>>\frac{\delta D_l}{\chi_s}$ and $\delta\sim Y$, the maximum growth rate occurs at the wavelength

$$\lambda=\frac{2\pi(\chi_l+\chi_s)}{\varepsilon\Lambda}(1+\Lambda\Gamma). \tag{4.25}$$

It can be thus concluded that the relation (4.25) fully coincides with that obtained in [19]. Therefore, in the short-wave approximation, with respect to diffusion members only, the calculation of the wavelength at which the growth rate is maximum is reduced to the plane problem.

In the second case, we neglect the diffusion member in (4.12), and it takes the form

$$\begin{aligned} &\frac{\partial T_{l1}}{\partial t} - V_s\frac{\partial T_l}{\partial r} - V_1G_l = 0, \quad \frac{\partial T_{s1}}{\partial t} - V_s\frac{\partial T_{s1}}{\partial r} - V_1G_s = 0, \\ &\frac{\partial C_{l1}}{\partial t} - V_s\frac{\partial C_{l1}}{\partial r} - V_1G_c = 0. \end{aligned} \tag{4.26}$$

Stationary solutions are constant for temperatures and concentrations. Relations of disturbance amplitudes (4.16) are then written as

$$\frac{\mathrm{d}T_{m1}}{\mathrm{d}r} - \frac{\omega}{V_s}T_{m1} = 0, \ \frac{\mathrm{d}T_{m2}}{\mathrm{d}r} - \frac{\omega}{V_s}T_{m2} = 0, \ \frac{\mathrm{d}C_m}{\mathrm{d}r} - \frac{\omega}{V_s}C_m = 0. \tag{4.27}$$

Therefore,

$$\begin{aligned} &T_{m1} = A_1\exp\left(\frac{\omega r}{V_s}\right), \ T_{m2} = A_2\exp\left(\frac{\omega r}{V_s}\right), \\ &C_m = A_3\exp\left(\frac{\omega r}{V_s}\right). \end{aligned} \tag{4.28}$$

Substituting (4.28) into boundary conditions (4.17), the following transformations with respect to (4.18) yield the dispersion equation:

$$\left(\frac{(\chi_s - \chi_l)\omega}{V_s} - \varepsilon\theta\right)\left(\frac{D_l\omega}{V_s} - (k_s - 1)(V_s + C_0\gamma)\right) - \gamma\varepsilon(k_s - 1)C_0\theta = 0. \tag{4.29}$$

Similar to the previous case, let us neglect members in (4.29), which contain $(k_s - 1)$ and obtain

$$\frac{(\chi_s - \chi_l)\omega}{V_s} - \varepsilon\theta = 0. \tag{4.30}$$

The maximum growth rate of disturbances is observed at the wavelength

$$\lambda = \frac{2\pi\sqrt{\varepsilon V_s\Lambda(\chi_s - \chi_l)(1 + \Gamma\Lambda)}}{\varepsilon\Lambda V_s}. \tag{4.31}$$

At the normal crystal growth, the rate of crystallization front is proportional to undercooling: $V_s = h\Delta T$ [19, 20]. Therefore, $\Lambda = h$, where h is the proportionality

Table 4.2 Parameters of the iron–tungsten system

Liquidus temperature T_{Lr}	1806 K
Heat of melting ΔH	270 kJ/kg
Density ρ_1	6980 kg/m^3
Diffusion coefficient D_r	10^{-8}m^2/s
Specific heat capacity c_0	611 J/(kg · K)
Thermal conductivity of liquid phase χ_1	6.8×10^{-6} m^2/s
Thermal conductivity of solid phase χ_2	6.9×10^{-6} m^2/s
Surface tension γ	1.788 N/m

factor, and ΔT is the undercooling. According to [20], the proportionality factor is calculated as

$$h_r = \frac{\beta D_{lr} \Delta HM}{\Delta lRT_{rL}^2}, \tag{4.32}$$

where M is the molar mass, R is the gas constant, Δl is the disturbance amplitude at the interface (~ 0.1 nm), T_{rL} is the liquidus temperature, and β is the coefficient considering that the free path of molecules in the liquid phase differs from the crystal lattice period and molecule symmetry. For symmetrical molecules, β ~ 10. According to Table 4.2 and (4.32), the value h_r is 0.558 m/(s · K) at the initial nucleus size 1 μm.

A transfer to the dimensional variable in (4.19) yields

$$\lambda_r = \frac{2\pi(\chi_{lr} + \chi_{sr})}{(\Delta H/c_0)h_r}\left(1 + h_r\frac{\Gamma_r}{\chi_0}\right). \tag{4.33}$$

The crystallization front rate is calculated from the previous paragraph: $V_{sr} = \frac{\Delta a}{\Delta t}$. Knowing undercooling, h_r is obtained as follows: $\Delta a = 10^{-8}$ m and $\Delta\tau = 4.4118$ ns, then $V_{sr} = 2.27$ m/s and $h_r = 0.757$ m/(s · K) at 3 K undercooling. The coefficient $\Gamma_r = \frac{\gamma T_{Lr}}{\Delta H\rho} = 1.71 \times 10^{-6}$ Km. According to (4.33), $\lambda = 0.382$ μm at $h_r = 0.58$ m/(s · K). At $h_r = 0.757$ m/(s · K), (4.33) yields $\lambda = 0.291$ μm. According to (4.31), $\lambda = 0.324$ μm at $h_r = 0.558$ m/(s · K) and at $h_r = 0.757$ m/(s · K), $\lambda = 0.242$ μm.

A comparison of the obtained data with the grain size in Fig. 4.5b, shows that convective and diffusion approximations explain the formation of grains by the normal growth up to 5 μm, although the maximum size is 10–15 μm. Hence, in the conditions of the normal growth, the model is not adequate.

Let us consider the growth caused by screw dislocations. The crystallization front rate is proportional to the square undercooling. Then, $\Lambda = 2\sqrt{hV_s}$ [19]. In this case, (4.25) takes the form

$$\lambda = \frac{\pi(\chi_l + \chi_s)}{\varepsilon\sqrt{hV_s}}\left(1 + 2\sqrt{hV_s}\Gamma\right). \tag{4.34}$$

Let us use dimensional variables in (4.34):

$$\lambda_r = \frac{\pi(\chi_{lr} + \chi_{sr})}{(\Delta H/c_0)\sqrt{h_r V_{sr}}}\left(1 + 2\sqrt{h_r V_{sr}}\frac{\Gamma_r}{\chi_0}\right). \tag{4.35}$$

For the growth mechanism due to screw dislocations, the kinetic coefficient h_r is calculated from

$$h_r = \frac{\beta(1 + 2g^{1/2})D_{lr}(\Delta HM)^2}{4\pi g R T_{rL}^3 \gamma V_m}, \tag{4.36}$$

where $g = 2\pi^4 n^3 \exp(-\pi^2 n/2)$, n is the number of molecular layers [25], and V_{m} is the molar volume. For metallic materials, $n \sim 6$, therefore, $g \sim 5.99 \times 10^{-9}$. The kinetic coefficient $h_r = 433$ m/(s · K^2), which is calculated from Table 4.2 for the model of the growth caused by screw dislocations. The wavelength is 14.8 μm, which is calculated from (4.30) at this kinetic coefficient and 10^{-7} m/s growth rate. This value coincides with the most probable grain size from Fig. 4.5b. In the convective approximation, (4.25) calculations show that the wavelength is of the order of 10^4 μm. Therefore, diffusional processes dominate at these rates of the crystallization front. When the rate is increased by three orders, $\lambda = 14.9$ μm that coincides with the experimental data. Thus, convective approximation is important at crystallization rates higher than 10^{-4} m/s.

Based on the above, it can be concluded that the Mullins–Sekerka instability adequately explains the formation of ~ 10 μm cellular structures at $V_{\mathrm{s}} < 1$ m/s and ~ 10^{-5} K undercooling. The volumetric heat source provides 10^{-7}–10^3 m/s rate of the crystallization front, depending on the power density of the heat source, which affects the material surface according to results from [26, 27] obtained by the phase field method, and the medium parameters. According to [28, 29], the effect of the heat source leads to large temperature gradients in the surface layers and, consequently, thermocapillary instability. Hence, in order to create the model of crystallization, it is advisable to consider instabilities differing from the Mullins–Sekerka instability, namely thermocapillary instability and concentration-capillary, which occur in the molten material. The analysis of dispersion equations obtained in [28, 29] for these instabilities shows that for the iron–tungsten system, the wavelength of the maximum perturbations is 12 μm that also coincides with the experimental data.

4.4 Conclusions

1. The solution of kinetic equations of the interface boundaries showed that the boundary R_2 of the liquidus temperature moved by the descending parabola, rather than by the dependence $R \sim t^{1/2}$, while the boundary R_1 of the solidus temperature moved linearly. At a certain time point, these boundaries approached to each other, indicating to a stopping of the crystallization process or the development of the crystallization front instability.

2. A further development of the crystallization model toward its adaptation to the plasma surfacing process of cast rolls included the rotation of one of the media and more strict account for the content of alloying elements.
3. Based on the linear analysis, the morphological instability of the crystallization front (the Mullins–Sekerka instability) was studied in two limit cases. In the first case, the convective member in equations was neglected, while in the second case, the diffusion member was neglected. It was shown that the mechanism of the normal crystal growth adequately explained the formation of cells not larger than 5 μm. This mechanism based on screw dislocations led to $\lambda = 14.8$ μm wavelength that coincided with the experimental data, if the crystallization rate was less than 1 m/s and undercooling was ~ 10^{-5} K. A further model development takes into consideration thermocapillary and concentration-capillary effects.

References

1. I.V. Mozgovoi, E.A. Shneider, in *High-Speed Steel Surfacing* (Omsk, 2016). (in Russian)
2. J. Damon, P. Schüßler, F. Mühl, S. Dietrich, V. Schulze, Short-time induction heat treatment of high-speed steel AISI M2: laboratory proof of concept and application-related component tests. Mater. Design **230**, 111991 (2023). https://doi.org/10.1016/j.matdes.2023.111991
3. G.-W. Park, S. Shin, J.-Y. Kim, Y.-M. Koo, W. Lee, K.-A. Lee, S.S. Park, J.B. Jeon, Analysis of solidification microstructure and cracking mechanism of a matrix high-speed steel deposited using directed-energy deposition. J. Alloys Compd. **907**, 164523 (2022). https://doi.org/10.1016/j.jallcom.2022.164523
4. N.N. Malushin, D.A. Romanov, A.P. Kovalev, V.L. Osetkovskii, L.P. Bashchen-ko, Structural-phase state of a high-hardness heat-resistant alloy formed by plasma cladding in a nitrogen atmosphere and high-temperature tempering. Russ. Phys. J. **62**(10), 1865–1870 (2020)
5. D.V. Aleksandrov, I.V. Aleksandrova, A.A. Ivanov, A.P. Malygin, I.G. Nizovtseva, Nonlinear analysis of solidification stability with phase transition. Rasplavy **2**, 27–43 (2014)
6. D.V. Aleksandrov, I.V. Aleksandrova, A.A. Ivanov, I.O. Starodumov, L.V. Toropova, Directional solidification with a two-phase zone with respect to the dependence of liquid phase density on temperature and impurity content. Rasplavy **1**, 37–45 (2020)
7. D.V. Alexandrov, L.V. Toropova, The role of incoming flow on crystallization of undercooled liquids with a two-phase layer. Sci. Rep. **12**, 17857 (2022). https://doi.org/10.1038/s41598-022-22786-w
8. M.G. Worster, Natural convection in a mushy layer. J. Fluid Mech. **224**, 335–359 (1991)
9. L. Arka, C. Abhik, Theoretical and numerical investigation of diffusive instabilities in multi-component alloys. J. Cryst. Growth **459**, 1–12 (2017). https://doi.org/10.1016/j.jcrysgro.2016.11.046
10. R.F. Sekerka, Morphological stability. J. Cryst. Growth **3–4**, 71–81 (1968)
11. D.V. Alexandrov, Solidification with a quasi-equilibrium mushy region: analytical solution of nonlinear model. J. Cryst. Growth **222**, 816–821 (2001)
12. A. Elsaid, S.M. Helal, Moving Taylor series for solving one-dimensional one-phase Stefan problem. Alex. Eng. J. **61**, 7121–7128 (2022). https://doi.org/10.1016/j.aej.2021.12.055
13. A.V. Lykov, *Theory of Heat Conduction* (Vysshaya shkola, Moscow, 1967). (in Russian)
14. S.A. Sergeev, Mathematical simulation of non-stationary heat conduction of crystallization of spherical particle in the melt with moving boundaries. Issledovano v Rossii 664–672 (2003)
15. M.W. Chen, Z.D. Wang, The evolution and morphological stability of a particle in a binary alloy melt. J. Cryst. Growth **607**, 127113 (2023). https://doi.org/10.1016/j.jcrysgro.2023.127113

16. G. Gildas, C.-A. Gandin, Morphological instability of a solid sphere of dilute ternary alloy growing by diffusion from its melt. J. Cryst. Growth. **448**, 17–20 (2016). https://doi.org/10.1016/j.jcrysgro.2016.03.041
17. G. Gildas, C.-A. Gandin, Analytical model for equiaxed globular solidification in multicomponent alloys. Acta Mater. **97**, 419–434 (2015). https://doi.org/10.1016/j.actamat.2015.04.030
18. J. Colin, P.W. Voorhees, Morphological instability of a solid sphere of dilute ternary alloy growing by diffusion from its melt. J. Cryst. Growth **448**, 17–20 (2016)
19. A.P. Gus'kov, Dependence of the structure period on the interface velocity upon eutectic solidification. Tech. Phys. **48**(5), 569–575 (2003)
20. A.P. Gus'kov, A.D. Orlov, Dependence of period of macrostructures on kinetic parameters under directed crystallization. Comput. Mater. Sci. **24**, 93–98 (2002)
21. V.F. Formalev, L.N. Rabinskii, On the Stefan-type problem with two unsteadily moving boundaries of phase transformations. Izvestiya RAN. Ser. Energ. **4**, 74–81 (2014)
22. E.M. Kartashov, Analytical methods for solving boundary value problems of unsteady heat transfer in regions with moving boundaries. Izvestiya RAN. Ser. Energ. **5**, 3–34 (1999)
23. H. Carslaw, J. Jaeger, in *Conduction of Heat in Solids* (Russian translation). (Nauka, Moscow, 1964)
24. M.-W. Chen, Y. Wang, H. Guo, The effect of anisotropic surface tension on interfacial evolution of a particle in the binary alloy melt. J. Cryst. Growth **510**, 32–39 (2019). https://doi.org/10.1016/j.jcrysgro.2018.12.032
25. J.W. Cahn, W.B. Hillig, G.W. Sears, The molecular mechanism of solidification. Phys. Usp. **91**, 691–719 (1967)
26. E. E. Slyadnikov, I.Y. Turchanovskii, Order parameter and kinetics of non-equilibrium phase transition stimulated by the impact of volumetric heat source. Russ. Phys. J. **59**(9), 1466–1474 (2017)
27. E.E. Slyadnikov, Y.A. Khon, P.P. Kaminskii, I.Y. Turchanovskii, Kinetics of nonequilibrium melting of a macrosystem under the action of a volume heat source. J. Eng. Phys. Thermophys. **93**(2), 389–400 (2020)
28. S. Nevskii, V. Sarychev, S. Konovalov, A. Granovskii, V. Gromov, Formation mechanism of micro- and nanocrystalline surface layers in titanium and aluminum alloys in electron beam irradiation. Metals **10**(10), 1399 (2020)
29. S.A. Nevskii, Formation of micro- and nanocrystalline surface layers of titanium and aluminum alloys during electron beam processing. Fundamental'nye problemy sovremennogo materialoved **17**(10), 385–395 (2020)

Conclusions

The aim of this monograph was to study the physical nature and formation of the structure, phase composition, and properties of the layer obtained by plasma surfacing in nitrogen and modified by high-temperature tempering and electron beam processing.

The data bank was created for the formation of the structure, phase composition, defect substructure, tribological and mechanical properties of the plasma-surfaced layer deposited in the shielding-doping nitrogen atmosphere onto medium-carbon steel modified then by multiple tempering and electron beam processing.

The obtained results are fundamental for the development of condensed state physics, general metallurgy, and thermal treatment and will promote the development of the theory of structure and phase transformations in high-speed steels due to electron beam processing. The understanding of the nature of the surface hardening of parts and mechanisms followed by tempering and electron beam irradiation is important for research into condensed state physics.

Research Findings

Research was conducted into the structure, phase composition, and defect substructure of the plasma-surfaced 1.3355 layer in the initial state and after high-temperature and electron beam modifications. This studies also concerned mechanical and tribological properties of the plasma-surfaced layer after tempering and electron beam processing.

The carbide network modification of the plasma-surfaced layer was investigated by TEM after tempering and electron beam processing. The analysis of the main factors leading to the structure and phase transformations in the ~50-μm thick plasma-surfaced layer after electron beam processing showed the increased microhardness, wear resistance, and reduced friction coefficient.

A. I. Potekaev et al., *High-Speed Steel After Tempering and Electron Beam Processing*, SpringerBriefs in Applied Sciences and Technology,
https://doi.org/10.1007/978-981-95-1398-7

The mechanism of the HSS solidification during plasma surfacing included the initiation and development of the Mullins–Sekerka instability at the crystal–melt interface. Dependences were suggested for the wavelength matching the largest perturbation mode, on the surface energy and undercooling.

MIX
Papier aus verantwortungsvollen Quellen
Paper from responsible sources
FSC® C105338

If you have any concerns about our products,
you can contact us on
ProductSafety@springernature.com

In case Publisher is established outside the EU,
the EU authorized representative is:
Springer Nature Customer Service Center GmbH
Europaplatz 3, 69115 Heidelberg, Germany

Printed by Libri Plureos GmbH
in Hamburg, Germany